Keep The

Bathwater

Emergence of the Sacred in Science & Religion

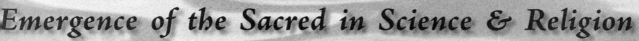

Best wishes in exploring the sacred "bathwater"

Ed Olson

5/23/09

Published by
Island Sound Press
Estero, Florida

Edwin E. Olson, Ph.D.

Foreword by Michael Morwood

Published by:

Island Sound Press

Email: edolson@islandsoundpress.com
Website: www.IslandSoundPress.com

Available at www.IslandSoundPress.com or:

Center for Sacred Unity
19691 Cypress View Drive
Fort Myers, FL 33967

Phone: 239-267-3525
Fax: 239-267-3043
Email: office@LambofGodChurch.net

Printed in the United States of America

ISBN: 978-0-615-27520-8

Library of Congress Control Number: 2009922047

Contents

Foreword

Jesus of Nazareth was an outstanding "light to the world" not because he brought God's presence to the world or regained access to that Presence, but because he opened people's eyes and minds and hearts to the reality of that Presence in their lives, however mundane and ordinary their lives might be.

The heart of his religious insight was simple and clear: living *in* love and living *in* God always go hand in hand. He invited people to name and affirm their everyday living and loving as intimate connection with the Divine – and he challenged people to give expression to this Presence.

Two thousand years later our rapidly expanding knowledge about our universe and how everything in it is interconnected invites us to name and affirm that living *in* the universe and living *in* God go hand in hand. There are no exceptions to this simple and clear insight. Everything that exists is sustained, energized, held in existence and connected by … what?

Scientists do their best to discover more about the "what". The more we learn from them the more we find ourselves in awe and wonder about ourselves as a life-form giving the universe a way to reflect on itself.

Religion now finds itself reflecting on the "what" also. It must do so because the "what", like "love" will be a significant pointer to the mystery that religion calls "God". Religion will never be able to describe "God" adequately, but hopefully in the 21st century religion will honor "God" not as an elsewhere, overseeing deity, but as a Presence permeating everything that exists. No exceptions.

We find ourselves, then, in the midst of the most significant shift ever in Christian history, a shift that challenges us to explore and almost certainly to change the concept of "God" that grounds Christian doctrine and that has traditionally been used to identify us as "Christian".

We need good guides and good tools to help us in this exploration. Edwin Olson is a thorough, gentle, perceptive and challenging guide. *Keep the Bathwater: Emergence of the Sacred in Science & Religion* provides the best tools that readers will find anywhere to help them to explore, reflect on, discuss, and synthesize the contemporary engagement of science and religion.

Michael Morwood
Balcatta, Australia
January, 2009

Dedication

To Greta Solia Olson, Tuesday Lili Hadden, Zeida Mueller Olson, Cisco Jean Hadden, and Clover Ann Hadden, my five granddaughters who have been born since my first book in 2001 which honored my children, their spouses and my three grandsons Preston, Arne, and Arlo Olson. May they and their children's' children's' children live their lives in a world of life-affirming adaptive systems that sees the sacredness in everyone and the planet.

Ed (Grandpa) Olson
Estero, Florida
January 2009

Preface

What's Needed?

In the midst of the many problems besetting us – the economic meltdown, ecological disasters, wars and terrorism, and breakdown of human institutions – an emerging consensus in science and religion is hopeful. Replacing outmoded dogmas and mindsets that fuel our global crises is a new understanding about the sacredness of the origin and connectivity of everything on the planet.

The goal of this book is to raise our collective awareness of this emerging consensus about the sacred. By increasing our knowledge and skills for relating to the sacred that is in us and everywhere around us we can rethink our purpose on earth, our care-taking of the planet, and our treatment of our fellow humans and other living things. This in turn will help us to be more resilient in dealing with crises, uncertainty, difference, and change.

Sacred Awareness

Sacred awareness is the capacity for recognizing and engaging the sacred in orienting our daily lives. With sacred awareness we can examine what is important to us about the sacred, including all of life and the planet; understand the information and experiences we use in interacting with others about the sacred; and examine the assumptions we make about the sacred that either perpetuate our separation or enhance our connection to others.

The components of sacred awareness addressed in this book are:

- Knowledge of alternative approaches to and definitions of the sacred from both science and religion (**Chapters 1-3**)
- Knowledge of principles and methods to explore the mystery and creativity of the sacred (**Chapters 4-6**)
- A coherent personal sense of the sacred and the importance of the sacred in our lives (**Chapter 7**)
- Competence in constructive dialogue with persons from different faith and secular traditions (**Chapter 8**)
- Commitment to address global issues in sustaining our sacred planet (**Chapter 9**)
- Capability of providing leadership about the sacred in developing organizations and systems (**Chapter 10**)

Complexity Theory

Complexity theory challenges the long-established worldview based on Newtonian physics that held that everything can be predicted or explained by the behavior of particles in motion. This reductionist view of the universe as one huge machine did allow us to develop scientific methods that provide understanding and control of much of nature, but does not address ethics, aesthetics, and free will.

Complexity theory looks across the sciences of cellular and evolutionary biology, ecology, mathematics, physics, information technology, psychology, and the the behavior of complex social systems, such as global economics, and demonstrates that all complex systems display properties that are non-linear and emergent. Elaborate patterns arise out of many simple interactions. These patterns become our values, meanings, morality, and even our religion.

These systems are fundamentally self-organizing, sustaining, and adapting. At each level of complexity entirely new properties apppear. For example, a chemist cannot understand the actions of chemical compounds by only knowing the behavior of elementary particles. How do ideas emerge from the firing of the 100 billion neurons in our brains when ideas are qualitatively different from neurons? Ideas consist of non-material meanings that cannot be reduced to particles of matter in motion.

Complexity science helps us make sense of what is happening inside and outside of the systems (including belief systems) we are in and then choose appropriate concepts, tools, and techniques to make changes. (See Eoyang, 1999; Goldstein, 1994; Kelly & Allison, 1998; Olson & Eoyang, 2001; Petzinger, 1999; Stacey, Griffin & Shaw, 2000; Zimmerman, Lindberg, & Plsek, 1998). These authors use concepts and methods from theories of chaos, complex adaptive systems, nonlinear dynamics, and quantum theory to develop innovative models for change.

The models that will be applied in this book are taken from Olson & Eoyang (2001) and the work of the Human Systems Dynamics Institute (hsdinstitute.org).

Layman's Perspective: Parable of Mustard and Yeast

In this book I will use complexity theory insights and methods I have used in my work in human systems to explore the emergence of a similar understanding of the sacred in science and religion.

I am not a theologian. I did graduate from a Lutheran college (St. Olaf) and received a master's degree in pastoral counseling from a Jesuit college (Loyola), and I have read widely, but I am clearly a lay person. I enjoy engaging in dialogue with persons of all faiths -- not to convert, but to share my thinking about the sacred and learn how this fits with other understandings of the sacred.

I am not a physical scientist. My training is primarily in the social and behavioral sciences, particularly the "soft" side of those sciences. I enjoy learning about the exciting new discoveries in the "hard" sciences such as those made possible by the new proton collider in France and the ongoing DNA studies.

I am an applied social, behavioral, and complexity scientist, which means I use the theories and methods of complexity theory and social and behavioral science in practical ways to understand and influence human systems. In recent years I have applied these methods with some success in projects and workshops with faith communities.

As an example of what this approach can yield, I will briefly reflect on the Parables of the Yeast and the mustard seed. The parables are about Jesus' teaching about the reign of God compared to everyday life in first century Palestine – to a woman baking bread and to gardens beset with weeds (Williams, 2008a). Both the planting of the yeast in a barrel of flour and the planting of the mustard seed in a garden cause disruption. Both the yeast (which leavens the flour) and the mustard seed (which becomes a shrub) are invasive, noxious and corrupting to the established order. And just as a mustard shrub disrupts a garden, we might not always value the results at first. The reign of God is likened to events in the everyday that are hidden at first but then take over if the ground is ready.

Through a complexity lens, the parables are about moving a system to a state of disorder so that a new order can form – one that self-organizes around a principle. In the parables the principle is God's presence in our daily life. The implication is that to engage with the sacred we need to be able to embrace and deal with the disorder that we encounter in life.

Bathwater metaphor

As we grow in our awareness about the diverse perspectives of the sacred, some will fear they will lose God in the process. Raymo (2008) provides a very helpful metaphor with a twist on the baby and the bathwater story. Raymo suggests that "the bathwater" represents the "mind-stretching, jaw-dropping, in-your-face wonder of the universe itself, the Heraclitean mystery that hides in every rainbow, every snowflake, every living cell". Raymo says this creative agency is worthy of attention, reverence, thanksgiving, and praise, hence the title of this book – Keep the Bathwater! The bathwater is the essence of what faith is about.

"The baby" stands for the "cultural accretions, the anthropomorphisms, misplaced pieties, triumphalism, intolerance toward "infidels", supposed miracles, and supernatural imaginings that religious traditions have placed on the mysteries of the universe" (Raymo, 2008).

Raymo says -- toss out the baby but save the bathwater! In this book we will explore aspects of what Raymo wants to discard -- the notion of a personal God, God in our own image, God invested with human qualities – and other aspects of the "baby" we have valued. However, the question of what "baby" to save along with the bathwater, what "baby" to discard, and what "baby" needs yet to emerge from the bathwater is left to the reader. Babies do not stay babies – they grow and change every day.

How the Book is Organized

Part I, The Sacred Bathwater, discusses the convergence of scientific and religious images of God and the emerging new story from both science and religion. **Part II, Increasing Sacred Awareness,** uses complexity and emergence theory to provide methods for exploring our sacred awareness, **Part III, Application of Sacred Awareness Principles**, applies principles of sacred awareness to self-development, dialogue with others about the sacred, acting as a good steward of the planet, and providing spiritual leadership in organizations. **Part IV The Sacred Awareness Journey,** discusses how we explore, accommodate, adapt, transform, and co-evolve along the path of sacred awareness.

How to use this book

The book presents new insights from both the sciences and theology that can raise our awareness of how embracing the new story about the sacred can provide an intellectual understanding of our faith that is also spiritually and emotionally satisfying.

This book is a resource for individuals and workshop participants in their exploration of concepts of the sacred and the common ground among the various religious traditions and between religious faith and scientific inquiry.

The individual reader can revitalize his/her understanding of how the sacred is intertwined with our daily lives and vital for our self-development. Study groups can use the book to help build an alliance between secular, religious, and spiritual progressives to deal with the compelling interrelated issues of our times including environmental sustainability, social justice, and spiritual fulfillment.[1]

---

[1] These three issues are the focus of symposiums developed by The Pachamama Alliance. www.pachamama.org

Acknowledgements

The complexity theory methods are derived from my work with Glenda Eoyang in developing our book *Facilitating Organization Change: Lessons from Complexity science* (Olson and Eoyang, 2001) and from the work of the Human Systems Dynamics Institute.

I would like to thank those with whom I had early discussions about some of the concepts in this book. They are: Rev. Bob Alley, Jim Ballenthin, Cheryl Crosby, Darold Daudt, Rev. Loren Grage, Milly Johnson, Martha Limburg, Tim Marr, Pastor John Monson, Diane Ruebling, Pastor Michael Small, Fred Smith, Karna Rinke, Beth Trout, members of the Hackensack Union Congregational Church and Salem Lutheran Church (Longville, MN) discussion groups, and my valued colleagues in the SAG group: Dr. Arland Benson, Pastor Charles Colberg, Michael Connelly, Dr. Bill Elliott, Tom Fisher, and Dr. Jim Limburg.

Special thanks to the readers of early drafts for their helpful feedback. They are: Judy and Marcel Charland, Pastor Charles Colberg, Ken Kostial, Rev. Larry Nelson, Dr. Eric Olson, Judith Olson, Don Reynolds, Vic Rinke, Larry Voeller, Dr. Glenn Whitehouse, and Cecelia Ann Smith who provided helpful edits. Shane Hadden offered valuable insight for reorganizing and focusing the book. Pastor Becky Robbins-Penniman greatly enhanced the final draft with cogent suggestions.

Several meetings sponsored by the Process Theology group and the North Central Program in Science and Theology (NCPST), Minnesota Consortium of Theological Schools were very useful in the early stages of my thinking.

I greatly appreciate the support of the Center for Sacred Unity at the Lamb of God Church, especially George Braendle for founding the Center, Marilyn Bowman who alerted me to the bathwater metaphor by Chet Raymo, and Pastor Walter Fohs who helped clarify the major themes of the book.

I am very indebted to Michael Morwood for his encouragement of this effort and for writing the foreword.

Without the constant support and love from my wife Judith this book could not have been written. She is indeed sacred to me.

Sacred Awareness Survey

Readers may wish to complete the following survey about sacred awareness before reading the book to gauge how closely their current thinking and beliefs about the sacred match the perspective presented in this book.

Sacred Awareness Survey

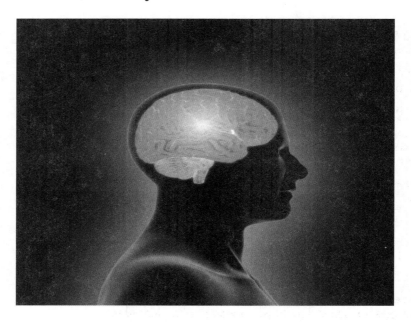

Focusing on the aspects of sacred awareness listed below, take the following 20-item survey. Be as candid as you can.

Item	QUESTION	Yes (✓)	No (✓)	Not Certain (✓)	Chapt.er Where discussed
1.	Are you able to identify what is sacred to you?				1-3
2.	Can you differentiate among the various views of the sacred in the world?				1-3
3.	Do you know why you believe the way you do about the sacred?				7
4.	Can you express your feelings, beliefs and thoughts about the sacred and defend them without alienating others?				8
5.	Is global warming a spiritual issue?				9
6.	Is being uncertain about your faith helpful in increasing your spiritual awareness?				5
7.	Would others who know you say you have a good sense of the diversity of views about the sacred?				1-3
8.	Is discussing what is sacred about work and organizational life an important component of leadership?				10
9.	Would you say your sense of the sacred contributes to a meaningful, rich and full life?				11
10.	Do you feel excited and satisfied with your engagements with others who have a different view of the sacred?				8
11.	Did people develop during eons of evolution rather than being added to the earth by a divine being after it was created?				1-3

Item	QUESTION	Yes	No	Not Certain	Chapter where discussed
12.	Are you free of dependency on others for understanding your approach to the sacred?				7
13.	Are your beliefs about science and your beliefs about religion compatible?				3
14.	Do you have an explanation for unexpected events that satisfies you?				4
15.	Do we as individuals and as a society have an obligation to reduce our impact on the environment?				9
16.	Do you know why your awareness of the sacred varies from time to time?				4-6
17.	Do you believe that the planet's biosphere is sacred?				9
18.	Do you provide leadership in your faith community about your sense of the sacred?				10
19.	Are social justice issues related to what you think is sacred?				8
20.	If a leader of a secular organization responded "yes" to most of the previous questions, would you encourage the leader to express his/her views in the workplace?				10

Interpretation of Survey

The 20 questions represent aspects of sacred awareness discussed in the book. The number in the right hand column refers to the chapter(s) in which the aspect is discussed.

If you had 15-20 "yes" responses, the material in the book will largely confirm your perspective on the sacred. If you had 10-14 "yes" responses, the book will provide a lot of new material for your consideration.

If you had 5-9 "yes" responses, much of the discussion in the book will stretch your views of the sacred. If you had 0-4 "yes responses, either the topics in the book could be very useful for you or not helpful for your faith journey at this time.

Sacred awareness is the capacity and willingness to understand the domain where both religion and science begin: what is life? Why are we here? Sacks (2007) says:

> Religion and science are like the two hemispheres of the brain, one analytical, the other integrative, one speaking prose, the other poetry. Religion without science is blind to the workings of the world. Science without religion is deaf to the music of creation.

Science tells us how the world operates. Religion tells us how it ought to be.

Sacred awareness is about knowing the difference.

Part I The Sacred Bathwater

Everything that we experience as material reality is born in an invisible realm beyond space and time, a realm revealed by science to consist of energy and information. This invisible source of all that exists is not an empty void but the womb of creation itself. Something creates and organizes this energy. It turns the chaos of quantum soup into stars, galaxies, rain forest, human being, and our own thoughts, emotions, memories, and desires (Chopra, 2000).

As the quote from Chopra suggests, reality is essentially energy and information. The womb of creation is a quantum soup, a bathwater, from which we and everything else emerges. Williams (2008b) points out that God is in this soup and within us all "to be reexperienced and reconceptualized generation after generation.

The objective of **Part I The Sacred Bathwater** is to explore the characteristics of this primordial soup, this bathwater, first from the perspective of science **(Chapter 1)**, from the perspective of religion **(Chapter 2)** and how the emerging views from both science and religion can be integrated in one worldview **(Chapter 3).**

Chapter 1 The Bathwater in Science

Scientific Worldview and God

In the 17th Century, Bacon (1561-1626) saw science as a means for the conquest and taming of nature. Galileo (1564-1642) used mathematics and experiments to discover the facts of nature and set the stage for many discoveries and applications of science. Descartes (1596-1650) argued that the human body was simply a machine with a separate and superior rational thinking mind. Newton (1643-1727) viewed the universe as a gigantic machine like a clock and demonstrated the natural laws of motion and gravity **(Figure 1-1).**

Figure 1-1 Sir Isaac Newton

SIR ISAAC NEWTON.

OB.1727.

These men were the architects of the scientific worldview in which the parts of the universe operate like parts of a machine, mechanically and predictably, according to the laws of physics and the laws of gravity and motion. Measurements, controlled experiments and empirical evidence bolstered their claims. They believed that the intricate designs in nature indicate there must have been a designer.

Early scientists had no problem in believing in God, because nature revealed his handiwork. In fact, it was Newton's beliefs about the characteristics of God that led to his mechanistic paradigm about nature:

- If God is omniscient, then nature must be absolutely determined.

- If God is eternal and sovereign, then nature must be composed of indestructible particles and irreducible laws.

- If God is one, nature must be uniform.

This worldview fostered a belief in one set of natural and deterministic laws that promised predictability and certainty about how the universe works. The ideas of linearity, order, invariability, uniformity, reductionism, determinism, equilibrium, indivisibility, dualism, and hierarchy can be traced back to Newton's notion of the divine.

A belief emerged that science of his day would provide information about the nature of God. According to Clayton (2004a):

> Newton's laws seemed to account for the interactions of all bodies in the universe. Yet, as Newton realized, applying these laws required an ultimate, unchanging framework of "absolute space" and "absolute time" within which bodies moved. This framework could be located only within God as the eternal object of God's thought--or at least it could exist only with the concurrence of God's will and as a reflection of the divine nature. So Newton's laws, the greatest insight in the history of physics, appeared to communicate something of the nature of God.

For Newton and his successors, the bathwater was a set of natural and deterministic laws that would explain the nature of God.

This mechanistic worldview produced the discoveries and inventions of the Industrial Revolution and it became the norm for all scientific endeavors including the work of Locke, Freud, Marx, and Darwin. Darwin's assertion that natural selection is the explanation for the patterns and designs in nature moved some to believe that everything results from designed laws with the details, whether good or bad, left to the working out of chance.

Increasingly God became absent from this worldview as the modern era progressed, unless God existed in some other spiritual or supernatural world – a dualism that separates the material world from the spiritual world, the body from the soul, the creation from the creator. Evolutionary science, the accuracy of mathematical physics, the strong belief in reductionism, and the dominance of materialist assumptions made science-based theological speculations difficult and, in the eyes of many, impossible (Clayton, 2004b). Ultimately, God was separated from the bathwater.

Even today, many scientists believe that nothing has been discovered in nature that cannot be explained with traditional naturalistic scientific methods. Some scientists and many theologians who do believe there are significant gaps in our knowledge of the natural world claim an intelligent designer, a "god of the gaps", as an explanation for the development of the intricacies of the eye, for example.

Beyond Reductionism and Determinism

Einstein (1879-1955) and others in the last 100 years have demonstrated the inadequacies of the mechanistic worldview. Energy and matter are two forms of the same thing. Light is a mysterious form of energy. At the subatomic level there is a quantum vacuum in which up to 200 elementary particles emerge and disappear in patterns and relationships. These particles exist in atoms for only a fraction of a second as part of subatomic interactions with an opposite number of "antiparticles" of opposite charge. These discoveries and others point out that there is no final simplification for which scientists have always searched. Brian Swimme (1999) describes the base of the universe as seething with creativity. David Bohm (2002) says there is a creative vacuum which is invisible to us out of which things and events arise and present themselves to us. Mystery and the possibility of the Divine are back in the bathwater of the universe.

With the Hubble telescope, we found that we are in a galaxy that is one of 140 billion galaxies that are moving away from each other at an ever-increasing speed. Scientists are able to read backwards from the present time to a point in time and space when everything started expanding outwards – the Big Bang, 14 billion years ago. The steps in the evolution since the Big Bang have been chronicled from the explosion of pure energy to protons, electrons, atoms, molecules of hydrogen and helium, massive stars, carbon, planets, life, evolution, brains, and consciousness. This is the new cosmology. With modern cosmology we have begun to understand that the reductionist, deterministic science that produced our technological society cannot explain everything. More and more events are discovered that could not have been predicted using the assumptions of the "old" science. The mystery deepens.

God Working Through Laws of Nature

Davies (2004), a physicist and cosmologist, believes that God initially formed laws governing the evolution of the Universe that allowed for self-organizing processes to emerge. The physical laws that facilitate the evolution of matter and energy are indeterminate, meaning that all of the elements of nature had to combine in ways that evolved the universe. Consider that there are only about 100 chemical elements in the universe that have combined to create the diversity we experience. There are only four letters (G,A,T,C) in our genetic language and only a few dozen amino acids that have given rise to the human species and the biosphere we live in.

The chance, openness, freedom, and contingency that God bestowed on nature are essential for creativity. Without chance, including the unpredictable behavior of humans, the world would be a preprogrammed machine. Gingerich (2000), a theistic scientist, argues that a contingent universe is organized with purpose and direction but not necessarily with a total blueprint. Collins (2006) believes that the natural phenomena point toward the divine and argues that both scientific and spiritual perspectives are needed for understanding of what is both seen and unseen in order to enrich and enlighten the human experience.

These scientists believe God works through the laws of nature. God upholds creation by being present in the simple generative rules in the universe that give rise to complex patterns. As we use these rules and adapt to the impact of the unpredictable interactions with other agents, new creative outcomes emerge. The massive entanglement of all of the agents of creation produces interactions that are unpredictable and uncontrollable with unanticipated consequences. Life and consciousness emerge at the edge of chaos where innovation and novelty combine with coherence and cooperation (Clayton & Davies, 2006; Davies, 2007). Some of these processes, such as the formation of galaxies, stars, and planets, are obviously beyond our control. But others, such as the development of our families, our social and work groups, and the societies we live in, are susceptible to human influence.

Conclusions

Science has broadened its vision of the bathwater from a set of natural and deterministic laws and a mechanical view of nature that created a dualism that moved God out of the bathwater. In this past century there has been a return to the mysterious and unpredictable nature of the bathwater. There appear to be generative rules that give rise to the complex patterns we experience and some scientists believe that the Divine is evident in the interactions of all of the stuff in the bathwater.

Chapter 2 The Bathwater in Religion

There are many names that describe the sacred – the Divine, Spirit, God, Mystery, Cosmic Power, Presence, Universal Soul, the More, Ground of Being, and the Numinous are a few. In the Bible the sacred is referred to as a potter, cup of cool water, path, safe place, a rock, a burning bush, an eagle, a whirlwind, a father, a mother, a king.

As we saw in the previous chapter, "creativity" is a term that represents a scientific perspective about the sacred (S. Kauffman, 2008). Creativity also represents how many theologians describe God's manifestation in the universe (G. Kaufman, 2004).

The choice of whether to use "God language" or creativity language to refer to the sacred is up to the reader. Essentially all of the terms for the sacred refer to the same phenomena – the amazing and endless mystery in the Universe.

Religious World View and God

The concept of a supreme God developed widely in both early religions and early Babylonian and Greek natural philosophers, and astrologers (Padgett, 2007). According to Plato, the universal pattern of causes and motion in heaven and earth implies that there is a "supremely good soul that takes forethought for the universe and guides it along that path" (Laws X, 897c).

Christian, Jewish and Arabic scholars agreed with Aristotle that God created the whole cosmos and sustains the principles and laws of nature. Over time, God became to be viewed as One who was all powerful, all knowing, all good, and absolute.

In early human societies the sacred was identified and expressed in stories, symbols, rituals, sacred times, and sacred spaces. Ethical practices and priestly hierarchies were developed to honor the sacred. For these societies the sacred was evident in nature and in human expressions through art, music and dance. Paths to inner peacefulness and physical healings as well as human expressions of love and loss were connected to the sacred.

In the Bible, for example, Moses was told by God that the mountain he was on was holy ground. What made any place sacred was the presence of God. "Surely God is in this place and I did not know it!" Jacob said as he experienced God in a vision. In **Figure 2-1** a man prays while a little boy inserts a scrap of paper into a crack of the Western Wall in Jerusalem, an enduring symbol of a sacred place.

Figure 2-1 Western Wall in Jerusalem

Impact of the New Cosmology on Theology

The new cosmology suggests that God is all around us. The cosmic consciousness and growing sense of complexity and mystery has increased the speculation by theologians about the universe and our place in it. It has also produced a sense of fear and insecurity that cause some to retreat to fundamental and simplistic answers.

Some theologians are exploring new metaphors for our beliefs in God in efforts to communicate a new worldview. Rabbi Nelson (2005) suggests five metaphors:

1. *Big Bang*. This demonstrates the uniqueness, the immense power, tremendous creativity of God that creates time and space.

2. *Fractal*. We are images of God. The scaling and self-similarity of fractals provide an image of how we are connected to God

3. *Light.* Light is fast and never ages. Light is comprised of photons from the Big Bang that are the same age today as they were then. Time has no meaning for God.

4. *Gravity.* Gravity warps time and space and creates the black holes in the universe that remain hidden, cloaked, and unknowable.

5. ***Calabi-Yau shapes*** (**Figure 2-2**). The geometric shapes are exceedingly complex, existing in all known dimensions but also in dimensions of reality that we cannot experience or imagine. The equations of superstring theory calls for more than the three dimensions we are used to. Calabi-Yau shapes are six dimensional energetic patterns named after two mathematicians. Superstring theorist Brian Greene (2005) says, "You and I and everyone else would right now be surrounded by and filled with these little shapes; the ultramicroscopic fabric of the cosmos is embroidered with the richest of textures." The theory suggests that we live and move in nine dimensions. In this metaphor God is everywhere in ways we can scarcely imagine.

Figure 2-2 Calabi-Yau Shape

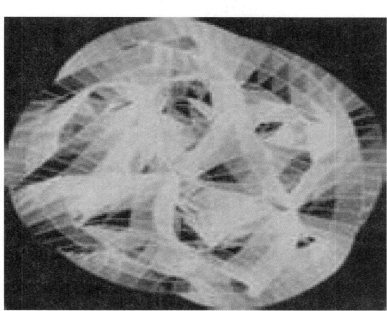

These metaphors suggest God is creative, the same in all parts of the universe, ageless, hidden, and complex. I would also add a sixth metaphor – **Initial Conditions**. We see in nature unpredictable events that are beyond our control, such as tornadoes and hurricanes. These phenomena form when the initial conditions are just right. The formation of these phenomena is sensitive to and depends on conditions being just right. This metaphor suggests that God and the sacred appear to us when the conditions make that possible. I will say more about creating conditions that reveal the sacred in Chapter 5.

Clayton (2004), a theologian influenced by the new cosmology, talks about the inherent tendency toward an increase in complexity, self-organization, and the production of emergent wholes that are more than the sum of their parts. New types of entities and new levels of complexity arise in ever new and different emergent realities. Clayton wonders what this might tell us about the existence and the nature of God.

Concepts of the Divine

Many religious scholars have integrated the implications of the new cosmology of the "Big Bang", evolution, and natural selection into their concepts of the divine.

Nasr (2000) says that there are four essential divine attributes: We come from God; we return to God; God resides inwardly at the center of our being; and the world itself is nothing but levels of divine presence.

Spong (2007) says,

> I experience God as life. I experience God as love. I experience God as being. I find God when I see life being lived, love expanding life and people finding the courage to be. I look inward to find the holy. I look at other people to see God at work in the power of love. I do not look to some distant far away place above the stars.

Gingerich (2000) says that:

> God may well have intended self-conscious, contemplative beings, quite possibly in many different latent possibilities fully known to the Creator, without specifying which of these possibilities would actually be realized…. This is a rather awesome thought. It means that maybe it is not in the terrestrial plan for everything to come out all right in the end. It means that we have some freedom to shape the destiny of human civilization, including both the freedom and the power to end it through greed, selfishness, and downright carelessness. This, then, is the implication of contingency.

Polkinghorne (2005) points out that God is at work in the flexibility of the creative process. He says:

> God does not fussily intervene to deliver us from all discomfort, but neither is he the impotent beholder of cosmic history. Patiently, subtly, with infinite respect for the creation with which he has to deal, he is at work within the flexibility of its process.

Although he acknowledges that God must be understood as ultimately mysterious, G. Kaufman (2004) states that God is creativity, and in and through creativity God is the ultimate source and ground of all our human realities, values, and meanings. Kaufman says:

> In the beginning was creativity, and creativity was with God, and creativity was God. All things came into being through the mystery of creativity; apart from creativity nothing would have come into being.

If God is present everywhere, or if everywhere and every thing is in God, it follows that anywhere and everything can be sacred. Process theologians such as Cobb (2007) point out that since God's work is always creative and always transformative, wherever creative transformation is occurring, God is there.

Sacred as Symbol

No one image or symbol captures the essence of the sacred. We connect to the sacred with our own personal images or symbols. These images and symbols are core to the stories of the sacred we create based on our experience. These symbols may be related to our recollections, our nostalgia, family, tradition, and culture.

For example, in a personal growth workshop the participants were asked to choose several symbols from an array of objects and talk to their small group about what the symbols meant to them. One man chose a small canoe with two paddles. He said that paddling the canoe represented a new journey. One of the paddles is to "keep me going in the right direction – my wife" and then he began to cry. The emotion came up suddenly and unexpectedly. For him that was a sacred moment **(Figure 2-3)**.

The canoe represented an aspect of his journey – every symbol he chose (a rock, cross, cream can, and tiger) was related to his journey. The story he told had a ripple effect on the group as they went deeper into their own stories. Surely this was a sacred space and the others knew it.

Figure 2-3 Couple Paddling a Canoe

Sacred as Fractal

A fractal is a geometric shape that can be split into parts, each of which is a reduced-size copy of the whole, a property called self-similarity. If we think of the sacred as a fractal, we can see patterns of the sacred everywhere. The canoe in the previous story was a fractal of the man's journey. All of the symbols he chose formed a pattern of his journey. We can see coherence in the patterns in nature, in the trees, and along the ocean shorelines. A great connection to the sacred is a fractal that is never completely finished. Wessels (2003) says that once we change from a human-centered view to an Earth-centered view, we experience the human community as sacred because of its relationship to the larger planetary community.

Our Need for the Sacred

Our need for the presence of the sacred is especially felt in times of trouble. We hope that the presence of the sacred will fortify us to deal with the situation. Given that the sacred is everywhere, the sacred is available and accessible to everyone. If we ask for it, we will receive it. If we seek it, we will find it. The sacred can satisfy our deepest longing to know that we are not alone.

From a process theology perspective, the Divine is expressed indirectly though gentle persuasion. We encounter the sacred here and now, not in the past or future. God/Spirit, the mystery of life, dwelling within us is constantly changing and evolving. God is incarnate in everything. Abraham Heschel (1976) says, " Not only does man need God, God is also in need of man. It is such knowledge that makes the soul of Israel immune to despair." What we do is the measure of the incarnation of God in the world. Living life from this perspective can be risky and messy as we stand up, speak up for the voiceless, and make personal sacrifices – but it can also be rich and fulfilling.

Peters (2008) writes that our relation to the sacred is a vital part of our identity. A connection with the sacred helps us find deep connections within ourselves – a sense of personal wholeness; connections with others and deepening of connections with the wider world – a feeling of being at home in the universe. Spiritual transformation involves making a change that puts the sacred as an object of significance in the life of the individual and a fundamental change in the pathways the individual takes to the sacred.

Peters also sees the sacred in the fundamental rhythms of our lives, e.g. weekly rhythms culminating in the Sabbath, monthly rhythms following the lunar cycle, annual rhythms of planting and harvest, or rhythms related to the equinoxes.

Bessler-Northcutt (2004), a professor of theology, believes that a new sense of the sacred is needed to relocate the primary experience of God to the midst of public life by developing a just and compassionate society. For example, an affluent community that neglects those living well below the poverty line must not really believe that the sacred is everywhere.

So What is the Sacred Bathwater?

The sacred is the creativity in the universe (science) and at the same moment it is God's presence in creation and in us (theology). We can take our pick or choose both. We can experience the sacred bathwater as:

1. A tangible place, time, object, or person in which we immerse ourselves to be refreshed and renewed.

2. Another reality beyond words and concepts, an awesome spiritual mystery that transcends time and space, the sense of a stupendous "More" or "Other". We can float in this bathwater and experience being one with the universe.

3. Emergent as we go about our daily lives. We are aware that the sacred is here and now, in images, in our breath, present in our daily work and play.

4. Patterns and rhythms of the universe in which we choose to participate. These patterns and rhythms repeat and are never completed. We choose who and what we hold to be sacred.

In any case, the sacred is what we value that gives meaning to our lives and our identity. The sacred helps us to:

- Embrace uncertainty
- Find deep conviction within ourselves
- Develop a sense of personal wholeness
- Make connections with others
- Deepen our connections with the wider world

Conclusions

Generally contemporary theologians believe that everything in the universe exists because it is continuously created by God. Denis (2006), for example, states that God creates in an emergent and evolutionary way in and through the whole process of the emergent universe.

Is God the creative activity or the ONE who is responsible for the creative activity? Are God and the creative activity one and the same? Is there a singular eternal reality, "One", a power, above, beneath, beyond and near all? These are questions that can't be resolved by either science or theology, but the discussion in these two chapters suggests that they both can be very accommodating to each other. The scientist can value religion's perspective on the mystery and divinity in what science studies. The new sciences are not trying to disprove theological concepts. Theologians can learn from science about new ways to image God. The creation story of modern cosmology can coexist peacefully with the creation stories of our religious traditions. The universe is one interconnected whole. Science and religion can be allies in increasing the level of spiritual fulfillment in the world.

Chapter 3 Integrating the Bathwater

The Common Water

The major point of agreement in the previous two chapters is that the sacred is everywhere. The sacred is the creativity in the universe (science) and at the same time it is God's presence in creation and in us (religion). Dowd (2008) says that although science and theology use different language, they are both talking about the same fecund process:

> To argue over whether it was God, evolution, or the self-organizing dynamics of emergent complexity that brought everything into existence is like debating whether it was me, my fingers tapping the keyboard, or the electrical synapses of my nervous sytem that produced this sentence.

Some scientists argue that the qualities of divinity we hold sacred – creativity, meaning, and purposeful action – are properties of the universe that can be understood and investigated scientifically.

Some theologians hold the concept that God is everywhere and have developed new metaphors for God such as Creativity. This chapter presents a contemporary understanding of the bathwater in which a new "baby" may emerge and play.

I will explore the bathwater from both a science and religious perspective. For this exploration, I will use the writings of three authors: a complexity scientist, a theologian, and a spiritual director who has integrated the new cosmology in her work.

Stuart Kauffman (2008), a pioneer in the field of complexity science and professor of biological sciences, physics, and astronomy, has published a new book, *Reinventing the sacred: A new view of science, reason, and religion.* He moves beyond reductionist science and believes that science has nothing to say about the meaning of human existence. His worldview is that there is a ceaseless natural creativity of the world that is a profound source of meaning and wonder. He sees a "divinity" in the universe without requiring it to be objectified in a personal God. If others want to attribute this creativity to a deity, he has no problems with that. His cosmology accommodates both theists and agnostics.

The dynamical change processes described by Kauffman argue that natural laws do not and can not explain everything. They can't explain novelty or human consciousness that is grasped by truth, unity, being, and beauty. The processes of emergence and creativity are altogether beyond the laws of physics. Complexity and other related sciences are needed to explain what happens when conditions are nonlinear and probabilistic.

The outcomes of the ceaseless creativity in the universe cannot be reduced to deterministic laws, according to Kauffman. Kauffman believes that life came naturally to exist in the universe, along with values, meaning, and consciousness with "God" as our chosen name for the ceaseless creativity in the natural universe, biosphere, and human cultures.

Science cannot provide answers to ultimate questions about why things exist and what their purpose is. At the root of the universe is a deeply mysterious unknown. Science cannot explain that or many aspects of human life such as love, friendship, and sacrifice. Science is a neutral way of explaining things, not anti-God or atheistic.

Michael Morwood, a former Catholic priest, has written *From sand to solid ground: Questions of faith for modern Catholics* (2007) that provides a theology that can be integrated with the scientific views of Kauffman. Morwood says the Spirit of God is the influence of that basic, sustaining mystery that holds everything in existence. From the planet's beginning, God's spirit has come to visibility in and through whatever was there, working in and through available thought patterns, knowledge, and worldview. This continued with the emergence of the human species.

All development took place in God, giving God a way of coming to expression in the magnificent life-form we are and could yet be. He believes all religions point us toward the mystery of God and the creativity in the universe that Kauffman describes.

Judy Cannato, a spiritual director and retreat director with a background in education and religious studies, has written *Radical amazement: Contemplative lessons from black holes, supernovas, and other wonders of the universe* (2006). She asks the question, "What does the new universe story ask of us?"

Cannato says we live and move and have our being in the midst of a Mystery that is deeper than ourselves and broader than our own creativity and genius can possible grasp. How we image God makes all the difference. God is a mystery, a Holy One who is concretely involved in every piece of life. All is in God (panentheism) in a dynamic endless state of creativity.

I will group the comments of these three writers somewhat irreverently into five categories related to the bathwater: preparing the bath, seeing what is in the bathwater, getting a big enough bathtub, creating waves in the bathtub, and being the "baby" by splashing and playing. We will return to these five categories in Chapter 7 as I develop principles for increasing sacred awareness.

See What is in the Bathwater

Water is not simply water – it is atoms of hydrogen and oxygen put together in a particular way that makes life itself possible. Raymo (2008) says water is a symbol for the creative agency that "wets the Earth with the stuff of life and consciousness – an agency worthy of attention, reverence, thanksgiving, praise."

Cannato emphasizes the fundamental connectedness in the universe and our need to be self-transcendent to evolve and connect with what is sacred in the bathwater.

Kauffman explicitly discusses the linkages between the natural universe, the biosphere, and human culture and encourages us to reinterpret our past concepts of the sacred and evolve our understanding. He makes the point that "new religions" have always built new temples upon the sacred sites of older religions and reinterpreted them. He urges that a scientific understanding of "the sacred" should be built upon earlier religious concepts, and not reject the traditions of the past out of hand.

Morwood emphasizes our relationship to our neighbor and the planet and urges us to deepen our awareness of what we stand for.

In **Part III**, this theme encourages us to connect to the significant differences that are all around us in the bathwater

Prepare the Bath

Baths do not just "happen", unlike sunsets they take intention and action to occur.

Cannato says that the new universe story asks us to expand our commitment to emergence and our contemplative awareness. We have the opportunity to nurture new life and use our listening heart to be open to what is possible.

Kauffman sees the ceaseless natural creativity in the bathwater and asks us to slow down to see the sacredness in the patterns in everyday life. Patterns are replicated with patterns from the last cycle informing the next cycle. The challenge is to step back and see the patterns and take time to reflect before acting.

Morwood (2004) offers a way of seeing and being in the presence of an "everywhere God" in his book *Praying a new story* (2004). He would see the bathwater as an opportunity to give human expression to the awesome mystery of God.

In **Part III**, this theme encourages us to participate in the emerging patterns in the bathwater.

Create Waves in the Bathtub

The waves in the bathtub disturb the sense of givenness, that the way things are is the way they always will be.

Cannato says the new universe story asks us to embrace a God of mystery, to be open to surprise and find relative comfort with the unknown. As we do, we will increase our capacity for both awe and taking action, to be fully aware that "something sacred is always afoot".

Kauffman says that with the dynamical change in the universe there is profound lawlessness, a creativity that is a profound source of wonder. This allows us to self-organize into more complex forms where development is an on-going process. As parts of the universe connect, feedback loops are created that produce new patterns. New complex forms and structures are formed depending on the length, width, and dynamic of the feedback loop.

Morwood describes God as the awesome, sustaining mystery that holds everything in existence. The mystery comes to expression in us. We are the life-forms "in which this mystery can express love, intelligence, joy, and delight. Here, on this planet, this mystery can speak and sing and dance and paint and create – because of the way of coming to visibility that you and I give it."

The uncertainty and lawlessness of creativity along with our self-expressions of the mystery cause waves in the bathwater. We need to create waves to fully engage the sacred.

In **Part III**, this theme encourages us to embrace uncertainty and move to a self-organizing space when possible.

Get a Big Enough Tub

You can be refreshed by drinking a cup of water, but you cannot bathe a whole baby in that same cup. You need a big enough vessel to do the job.

Cannato encourages us to expand our image of God to include a divinity that is both transcendent and immanent. The bathtub has to be at "least as big as life, as expansive as the universe". This will expand our capacity for personal empowerment.

Kauffman sees God in the effects of the universe where life, the biosphere, consciousness and the economy emerge from processes that are not reducible to physics. His sense of the bathtub is that it offers many opportunities for development of our species.

Morwood believes the biblical stories emerged from the Spirit of God working in and through the thought patterns, the knowledge available, the worldview at the time, and the questions with which people in that time long ago were struggling for answers. We need to create our own stories of the meaning of God to give understanding and hope in today's world.

God's Spirit can come to visibility if our "tub" is big enough for stories that are inclusive and relevant for our times.

In **Part III**, this theme encourages us to enlarge our boundaries to move into the adjacent possible

Be the Baby: Splash and Play!

A baby that loves bath time is finding joy in everyday life.

Cannato stresses the importance of communion and our capacity for agency – for using our unique gifts for the benefit of the whole.

Kaufmann says that, as co-creators, humanity contributes to the development of the universe in the direction of our own ideals and values. These patterns are replicated as smaller complex systems connect together to create larger forms.

Morwood says that the language of connectedness, relationship, presence, gift, and trust is the best language we have for connecting with God, but we need to be wary of constructing an image of God according to human notions of "person" and insisting that is what God is really like. He says we need to give less attention to who God is and focus more on the wonder of who we are and our capacity to contribute. Our reflection, our times of prayer, and our times of community worship should focus on deepening our self awareness about what we stand for.

In our interactions we can make a big splash and grow the baby to be what Cannato calls a "universal human" – integrated body, mind, and spirit.

In **Part III** this theme encourages us to use our talents to contribute to the whole.

Emergentist Theology

Emergentist theology looks forward: it postulates God as the goal toward which all things are heading.

According to Clayton (2004b) and Gregersen (2004), traditional theology looked backward: it postulated God at the beginning of the universe designing a universe intended to bring the world into existence as we know it today. For this to happen God would have to use deterministic laws to bring about the desired outcomes. To remedy any gaps, God would have to set aside the original laws in order to bring about a different outcome. Divine action then becomes the working of miracles, the breaking of laws; and God becomes the being whose nature and actions are opposed to nature.

Emergentist theology argues that we arose out of the very physical dynamism of the cosmos which is self-organizing and creative (Johnson, 2007). Because creativity is unpredictable, new and surprising events are constantly unfolding, but all are doing so within the divine bathwater.

Conclusion

Rabbi Sir Jonathan Sacks (2007), chief rabbi of the United Hebrew Congregations in England, wrote an article in the London Times arguing that any account of the human condition that reduces the human spirit to an accidental by-product of evolutionary pressures tells less than half the story of who we are. Sacks says:

> We may be — on this, the Bible and neo-Darwinism agree — "dust of the earth", the reconfigured debris of exploded stars. But within us is the breath of God. Scientists call this "emergence": the process whereby systems of self-organising complexity yield something new, more than the sum of its parts. That is where religion and science both began: when life became conscious, then self-conscious, then able to ask the question: "Why?"

From earliest times, as human consciousness developed, the sacred emerged from the bathwater of the universe as an answer to the question: "Why?".

Introduction to Chapter 4 Complexity and Emergence

This chapter discusses patterns and emergence in the ongoing creation. The chapter provides a general introduction to complexity theory, self-organizing processes, and the adjacent possible. The emerging butterfly in **Figure 3-1** suggests the process of emergence.

Figure 3-1 Emerging Butterfly

Part II Increasing Sacred Awareness

The new story of the integrated sacred bathwater in science and religion provides the basis and rationale for going more deeply into the concepts of emergence and complexity **(Chapter 4)**.

These concepts, in turn, provide a springboard into some specific complexity methods **(Chapter 5)** we will use in the remainder of the book to increase our sacred awareness.

In **Chapter 6** we apply these methods to examples of engaging with the sacred.

In **Part III Application of Sacred Awareness Principles** we look backward to **Sections I and II** to describe five simple principles we can use going forward into **Chapters 7-10.**

Chapter 4 Emergence and Complexity

Changing Assumptions

Many people are uncomfortable about thinking outside of the box of either their religious faith or modern science – for good reasons. A person's faith and scientific understanding provide comfort and security amidst the challenges of everyday life. Both modern science and religious practices have been productive in guiding human behavior and productivity. Why upset the apple cart and challenge our beliefs that there is a universal set of assumptions that supports both faith and reason? We want to believe that the world revolves around one universal set of assumptions and that the assumptions that guide our life and daily activity somehow fit together. This would seem to make sense. If there is one universe, there should be only one way to know the universe, however, we need two sets of assumptions to explain things.

The first set of assumptions is the familiar linear, deterministic, mechanistic laws of traditional science. They are very useful and can be used to predict many aspects of our life; however, as Kauffman (2008) argued in the previous chapter, they don't explain everything. The first set of assumptions does not lead to finding meaning in the universe. We need a second set to do that. The second set of assumptions recognizes that the world is emergent and self-organizing. Although we cannot control it, we can influence events. This second set of assumptions gives us reason to believe that there is sacredness to how things happen, even though at times we do not like the outcomes. We need to embrace both sets of assumptions as the amazing creativity in the bathwater of the universe pointed to by both scientists and theologians (Chapter 3).

Emergence and Complexity Theory

The term "emergent" was coined by the pioneer psychologist G. H. Lewes (1875), who stated that emergents cooperate by adding things of unlike kinds; the emergent is unlike its components and it cannot be reduced to their sum or their difference. Goldstein (1994) defines emergence as: "the arising of novel and coherent structures, patterns and properties".

The universe emerged in the first fraction of a second after the "Big Bang" **(Figure 4-1).** We do not know what happened at time zero (or before time zero). Cosmologists can give a reasonable scientific account of the emerging universe from the end of the first second when ordinary particles such as protons and neutrons and electrons existed. As the universe expanded and cooled, the nuclei of the simplest element, hydrogen, and some of the helium were formed. These nuclei in turn created an expanding and cooling fireball.

Figure 4-1   The Big Bang

After 400,000 years, the fireball was cool enough for nuclei to bond with electrons and form atoms of hydrogen and helium. We are remnants of the primordial fireball that carried within itself the potentiality for the universe – of all that would ever emerge.

Our Milky Way galaxy is emergent. Through the Hubble telescope we see supernovas and new clusters of galaxies emerging.

Figure 4-2   Andromeda Galaxy

The Andromeda Galaxy **(Figure 4-2)** observed by the Hubble telescope is the closest galaxy to our own Milky Way Galaxy -- 2.3 million light years from Earth.

Emergence happens in nature as local actions in different areas connect. When connected, local actions combine in unexpected and often sudden ways to create more complex systems. These new systems always have more capacities than the sum of their individual parts.

In nature, the exact shape and dimension of emerging patterns cannot be predicted **(Figure 4-3)**. No two termite mounds are alike. They arise as a result of the unpredictable interactions of the termites.

Figure 4-3 Giant Termite Mound in Outback

Emergence is not true of only galaxies and termite mounds; as part of the universe we, too, are emerging. We can resist this concept, or find ways to embrace it. Complexity scientists embraced it, developing complexity theory and complexity science to explain and work with the unpredictability of emergence. Complexity scientists focus on understanding how relationships, interactions, small experiments, and rules shape emerging patterns in unpredictable situations. When prediction and control are not possible, we need to adapt. Small changes right in front of us can morph into large-scale change, much like an avalanche where small beginnings escalate into dramatic and powerful forces. For example, with the connectivity of the Internet and the media, an email or picture of a breaking news story can reach millions within minutes.

Patterns

As I look outside my home I see and even hear patterns. The leaves in the trees form a pattern; the ripples on our lake are patterns; the raindrops I hear and see hitting the porch form patterns.

When I start a conversation with my neighbor "Sam", a pattern soon emerges. It could be a conversation about golf, the weather, or our travel plans for this year. If I focus on what Sam and I have in common, a familiar (and safe) pattern will emerge. If I move to what is different between us, such as political preference, a new (and perhaps uncomfortable) pattern will emerge.

Continuing the example, if Sam and I have a habit of engaging in "safe" conversations, it is difficult to break that pattern. Once a pattern is formed, it constrains our interactions. When we interact, we have mutual influence in building our relationship. We can both determine if we like the pattern that is forming or if it is getting boring and having a negative effect on those standing around us.

We can take this example of Sam and me and scale it up. Is the pattern of our collective consumption in North America putting such pressure on Central and South America for resources that we are contributing to the destruction of the rain forests and endangering life on the planet?

This is a big shift in discussing patterns, but the pattern that emerges from our interactions can be small and mundane or large and impactful. Patterns start from interactions that in the beginning appear unorganized but soon blossom to highly organized patterns that constrain or limit our choices.

Most of us have a vision of the life we want for ourselves and our families, but a vision is a pattern we can achieve only by emergence. What pattern emerges as we go through life is usually different from what we envisioned. The pattern may be even better than we imagine. The point is not to be so locked into one vision that we do not adapt to new possibilities. One pattern or choice may not be better than another.

Many patterns of thought and human interaction that we see about the sacred were formed by the experience of people many years ago. If we think of the metaphor of the baby and the bathwater, we can create new patterns that will be more meaningful for our new understanding of the bathwater.

Emergence's Engine - Self-Organizing

The processes that create new reality, novelty, and complexities are self-organizing. Religious scholars would say the processes are expressions of God's creativity. Scientists see self-organizing as a natural, albeit mysterious, process.

At the core of complexity science is the assumption that the richness and diversity in our evolving universe has emerged from self-organizing processes that produce order from the initial chaos. Some of these processes, such as the formation of galaxies, stars, and planets, are obviously beyond our control. But others, such as the development of our families, our social and work groups, and the societies we live in, are susceptible to human influence.

When we participate in self-organizing processes we create increasing complexity. It is like weaving a rug with an intricate design. When we follow and repeat simple rules, the rug design can become quite complex.

In self-organizing human systems, change occurs as the results of small local, personal interactions. Small changes by us, individual agents of creation, become integrated and scaled throughout larger systems and processes. The spontaneous movements of self-organizing, under the right conditions, produce order and stability in unorganized situations. Individuals and communities can create self-organizing processes that facilitate adaptation to changing conditions.

In **Figure 4-4** an anonymous group is self-organizing. The exact arrangement of the people and the outcome of the gathering cannot be predicted.

Figure 4-4 People Self-organizing

Figure 4-5 illustrates how individual agents, which could be people, things, ideas, or anything within a boundary, have the freedom to interact in unpredictable ways and whose actions are interconnected such that they produce patterns. The agents could be geese preparing for their journey south for the winter as individuals self-select in taking turns in leading the formation.

Figure 4-5

Self-Organizing

Individual agents, who are free to act in unpredictable ways, interconnect to produce system-wide patterns.

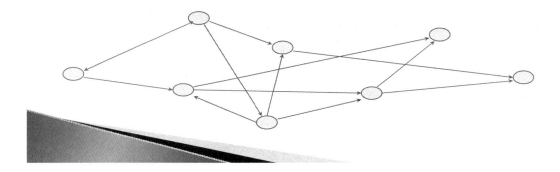

In **Figure 4-6** we see the patterns produced by the self-organizing agents. As agents interact, they inevitably produce a pattern. These patterns change as the interactions change, so many patterns are possible. Note that as the agents create or reinforce a pattern, the pattern also has a constraining effect on the interaction of the agents. Unless those interactions change, the same pattern will be reproduced. For example, we see this in organizing a parade. Each year there are some different floats or bands so each parade is unique, but the basic structure from previous successful parades is replicated.

In **Figure 4-6** the arrow on the left indicates that the parts create the whole, the pattern. The arrow on the right indicates that the pattern that has emerged affects the interaction of the parts. This phenomenon is often described as upward causation (from the parts to the whole) and downward causation (from the whole to the parts)

43

Figure 4-6

Formation of Patterns

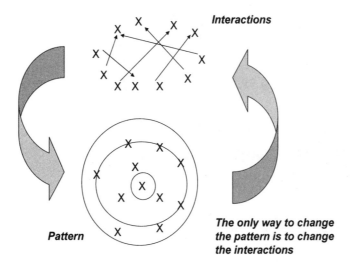

Engaging in a self-organizing process requires self-awareness, self-management, and courage to take risks and accept losses for an uncertain outcome. In most situations, outcomes are similar to what the philosopher Forrest Gump said about life: "Life is like a box of chocolates; you never know what you are going to get."

Adjacent Possible

What is the adjacent possible? The concept is easy to grasp if you think of playing dominos (**Figure 4-7**). In the game you move to an adjacent space that is possible, given the circumstances and the pieces you have to play.

Figure 4-7 Dominos

Kauffman (2008) speculates that the idea of the adjacent possible is one of the general laws of the universe. He believes that, as an imperative, biological species keep jumping into what is adjacent and what is possible. In so doing they increase the diversity of what can happen next. However, the movement into the adjacent possible needs to be paced. To move too fast would destroy a species' own internal organization. In other words they "explore the adjacent possible as fast as they can get away with it".

Applying this law to our own species, it is a useful metaphor for increasing our intelligence about the sacred. We are influenced by what we see and experience. Where we put ourselves and the actions we take in response to these new experiences affects our thinking. If we place ourselves in situations to learn other perspectives on the sacred, we can incorporate them into our way of thinking.

When we see emergence and make conscious choices about an adjacent opportunity we can be transformed. When we ignore or miss the emergent, the previously unseen adjacent opportunities are also missed. Each person needs to determine his or her own adjacent possible.

Conclusion

The emergence of novel and coherent patterns has been studied by complexity theorists and scientists. They have found that the processes that produce this emergent reality are self-organizing. Although self-organizing is still somewhat of a mystery, how emergence of new phenomena occurs in human systems is understood well enough to formulate methods for influencing the process. By affecting the interaction of the agents in a human system, we can change the resulting patterns. The emerged patterns cannot be exactly predicted or controlled because the interactions producing the patterns are nonlinear and probabilistic. We can use the uncertainty of emergence to spur us on to adaptive and creative action.

Accepting the reality of emergence and the assumptions that go with it challenges both religion and science. Two sets of assumptions are needed to explain events in the universe. One familiar set is linear and deterministic. The other set of assumptions is emergent and self-organizing. Understanding the sacred and developing sacred awareness requires us to embrace both sets of assumptions and embrace the constant creative process that the universe makes possible through all of our adjacent possibilities. Each day we take in the wonder of creation and we give out our time and attention to what matters to us.

Introduction to Chapter 5 Complexity Methods

To help step out of the "box" of our view of science and theology, we will provide metaphors and practical methods derived from complexity theory in Chapters 5. These metaphors and methods provide alternative ways of knowing and acting by understanding and influencing the forces around us rather than try to control the outcomes. The three methods are:

- **Conditions of Self-Organizing** -- Changing the speed, path, and outcomes of sacred unity by influencing relationships, containers, and significant differences.

- **Certainty Patterns** -- Identifying and impacting aspects of the sacred that are organized, self-organizing, and unorganized.

- **Stretch and Fold** -- Stretching into differences and concerns and folding to integrate and focus the energy.

Chapter 5 Complexity Methods

The creativity in the universe and the phenomena of emergence and self-organizing account for the changes we experience. We have seen that interactions by us, the agents of creation, produce changes that become integrated and scaled throughout larger systems and processes. Self-organizing, under the right conditions, produces order and stability in unorganized situations.

As one of the metaphors for God, **Initial Conditions** affects the unforeseen outcomes of emergence. We can do nothing to affect these initial conditions, but we can affect the speed, direction and outcomes of self-organizing process. At the largest scale, our collective actions affect global warming, melting of the glaciers, and depletion of the ozone layer. International conflicts are ignited by the behavior of nations and terrorist movements. Individuals can have an effect in communities, organizations, groups, and families.

In this chapter we identify the specific conditions we can influence that will move a system to a state of self-organizing. They are the three conditions of Containers, Significant Differences, and Transformative Exchanges. In reference to the sacred, these conditions speak to the questions below in **Figure 5-1**:

Figure 5-1 Conditions of Self-organizing and the Sacred

Condition	Question	Response
Containing	**Who am I in relation to the sacred?**	**My identity includes the sacred**
Differentiating	**What is important about the sacred?**	**The sacred is the source of my values**
Exchanging	**How do I engage the sacred?**	**The sacred is expressed in my relationships**

The Conditions of Self-Organizing

As agents interact and self-organize there are three conditions that will affect the outcomes of the self-organizing:

- The kind, size, and tightness of the **container(s)** in which the agents interact.

- The significant **difference(s)** that is identified that will give direction and shape to the emerged patterns.

- The quality, kind, depth, width, and duration of the **exchanges** between the agents.

These conditions are at the core of understanding how self-organizing works in the universe and in human systems. Regarding the sacred, the three conditions affect the images and expectations of the deities we may inherit or create. We can influence one or all three of these conditions to change the pattern of our lives and our engagement with the sacred.

The lived-out responses to these questions create the patterns of our spiritual life – our identity, our values, and our relationships. Understanding these three conditions helps us to not only sort out what the sacred means to us, but also to change our identity, values, and relationships to be more life-affirming. Changing the conditions changes the patterns of our life.

The three conditions affect the interaction of the agents in a self-organizing system, which in turn affects the emerging patterns. The patterns then influence the conditions, which then reinforce the patterns. The patterns stay stuck until the interaction of the agents changes one or more of the conditions. We will explore how this works with an example but first let's define the three conditions.

Containers

The containing condition involves our identity. The boundaries we set, the goals we have, and the rules we live by provide a container within which self-organizing occurs as we live our lives forward. We can think of the containing process as a magnet or a fence holding things together. Our cultural, racial, gender, and religious affiliations are containers that bind us together or separate and divide us. We are within multiple containers at the same time. These could be our family, neighborhood, faith community, sports team, age cohort, and career. The chalice in **Figure 5-2** is a container.

Figure 5-2 A Chalice as a Container

48

Significant Differences

We have many differences in our lives but only a few become significant and form the basis of a new pattern. These differences are usually what we value. In the universe the significant differences could be any number of polarities such as light and dark, hot and cold, or good and evil. In human systems they can be social issues, power differences, knowledge levels, or personal characteristics. Whatever difference emerges as most significant will shape the emergent patterns. Examples include roles, rules, beliefs, ethnic difference, social norms, geographic area, seniority, and schedules. The cute baby and puppy in **Figure 5-3** have wrinkles in common but there is a significant difference between a baby and a puppy.

Figure 5-3 Significant Difference – Baby and Shar-pei puppy

Transformative Exchange

When there is a meaningful exchange across the significant differences in our lives, transformation occurs. The exchanges could be the transferring of energy, information, or material resources. When we make contact with another agent and exchange something of value, our relationship changes and these exchanges generate new patterns for others. The interconnectivity creates compassion, a shared joy, shared suffering and a struggle for justice (Fox, 2006). Examples include love and support, communications, mentoring, learning, transfer and acceptance of responsibility, conflict resolution, meetings, delivery of services, and recognition. The picture in **Figure 5-4** shows a parent handing over the keys to a car to a teen-ager – an exchange likely to change the dynamics in the household.

Figure 5-4 Exchange: Handing Over the Keys

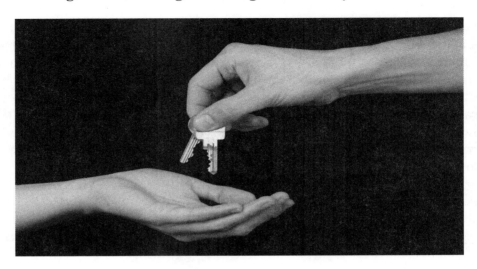

The Conditions in Early Religion

We can see the conditions for self-organizing in the questions that arose in all of the early religions. When people began to wonder about who they were and what were the mysterious forces that affected their lives, they were raising questions about **containers.** Was the sacred the sun, moon, the sea, a volcano, the wind? Whether the sacred was close in everyday things (animism), distant on far away mountains, or in the stars, were questions about the size of the boundaries. How these questions were answered affected the size, scope, and numbers of the deities they imagined.

Questions were asked about the nature of the sacred. What are the attributes of the deity? How is our god different from the gods of the tribe who has defeated us? Is the sacred good and loving, jealous and vengeful, all-powerful, mindful of our actions or inattentive? How do I worship this deity? Does god intervene regularly or occasionally? The answers to these questions were the significant **differences** that shaped the various emerging patterns. The assumption that there is an entity behind all of the creative and sustaining activity in the universe was itself a huge difference that shaped the concept and image of God. The differences then affected what was given attention and priority. Do we pray, do we offer sacrifices? Is there an afterlife? How do I prepare for it?

Questions also emerged for individuals about how they should relate to the sacred. What is expected of me? How do I give of myself to this god? What will I get in return? Do I serve this lord by showing compassion for my neighbor? Do I approach a sacred place with trust and gratitude or in fear? The transformative **exchanges** between the individual and the sacred affected their sense of well-being and the rules that guided their behaviors toward others.

Options for Influencing the Conditions of Self-Organizing

To affect a **container** in a human system, we can make it larger by adding more people, moving to a larger space, add more objectives, broaden the scope, or open the schedule. To make a container smaller, we can work with fewer people, move to a smaller space, reduce the scope, reduce the number of objectives, or tighten the schedule.

By identifying the **differences** that are significant we create new patterns, which could be a focus on common ground or a reduction in conflict. If we over focus on one differentiating aspect, the system will split (bifurcate) so that a difference creates a separate container and new patterns emerge.

To loosen **exchanges** we can take more time to respond, use printed communications, use analogies and stories, dialogue and reflect, and ask questions. To tighten exchanges we can respond quickly, use face-to-face communication, ask for return messages, be specific, and make demands.

To further explain the conditions of self-organizing, I will use an example of baking bread. The same examples will be used to illustrate the other two complexity methods.

Baking Bread

I built an outdoor brick wood-fired oven (**Figure 5-5**) because I was dissatisfied with the quality of the available bread from stores and bread produced by my bread machine. I believed that artisan whole-grain bread could be baked in brick ovens and still be tasty and healthy, and I was curious about the variety of breads that could be baked in this manner.

To get the desired artisan bread it was necessary to build an oven that produced high temperatures using all three modes of heat: convection (flowing hot air), conduction (contact with the hearth), and radiant (heat from the sides and top of the oven). Choosing the right kind of flour and yeast is crucial.

Figure 5-5 Wood-Fired Outdoor Brick Oven

To carry out a successful baking process, I needed to develop a timetable, follow the rules in measuring ingredients, mixing, kneading, and baking. To maintain a sourdough culture I had to feed it regularly – this turned out to be my biggest problem.

Container. The oven itself is a container of firebricks encased in concrete to return the heat with an insulating layer of vermiculite and a stucco shell. The wood fire is built on the hearth of firebricks and burned until a temperature of 700° is reached (for pizza) or 500° (for bread). A container is essential to reach and hold the high temperature until the baking is complete.

The bread dough itself requires a container (a bowl) as it is mixed, proofed, and formed into a loaf ready for baking. As it forms a "skin" the loaf will hold its shape and inherently uses a containing process.

Difference. The right ingredients are essential. Most any combination of water, flour, and salt will yield humdrum but eatable bread. By adding leaven and other ingredients, spectacular loaves will emerge. For example, by adding a starter of fermented yeast in the process, the loaf will explode with flavor and air pockets. Using more water and less flour in preparing the loaves, which is possible in a brick oven, produces bread that will stay moist for days. If you want to add raisins, olives, or well… you get the idea.

Exchange. The mixing of the ingredients through folding the dough back on itself, gently, to create layers of dough is essential. Letting the dough rise in a warm temperature to ferment the leaven is the next exchange. The final one is to place the dough on the hot hearth to form a crust and trap the moisture with intense heat inside. In each of the exchanges the ingredients react to and transform the others.

Since all three conditions are essential and interconnected, change will occur when a change is made in just one of them. Changing the heat (container) affects the quality of the crust; changing the ingredients (difference) will change the flavor; changing the way the ingredients are kneaded (exchange) will affect the density.

Certainty Patterns

As we encounter daily life, there are moments when we are certain of what will happen, things are organized, and we find we are in general agreement with what is happening. At such time we can confidently plan and have a sense we are in control.

In other instances, things seem unpredictable, unknown, or unorganized, and we do not like what is happening. At these times we feel very stretched and subject to the whim of random changes.

It is the space between these two extremes that offers the most possibilities for creative action with the sacred.

In human systems, we can create a continuum of certainty-uncertainty to diagnose when situations are amenable to self-organizing. In **Figure 5-6,** situations on the far left of the continuum with a high degree of certainty are organized, predictable, controlled, and orderly.

Figure 5-6 Certainty Continuum

Organized	Emergent	Unorganized
Predictable	Adapting	Unpredictable
Orderly	Changing	Random
Controlled		Uncontrolled

←---→

High Certainty　　　　**Self-Organizing**　　　　**Low Certainty**

Situations on the far right in the area of low certainty are unorganized, unpredictable, uncontrolled, and random. In between where there is both some certainty and some uncertainty, situations are emergent, adapting, and changing like a sand dune in a desert. This is the area of self-organizing. We can experience all three states at the same time in different spheres of our life.

What is happening in each of the three states is illustrated in **Figure 5-7.**

Figure 5-7 Behaviors along the Certainty Continuum

Agreement	Deciding	Disagreement
Clear expectations	Relationships	Unknown work
Directed action	Learning	Whisper of trends
Tight linkages	Growth	Unclear expectations
Heavy controls	Influencing	Loose connections

←---→

High Certainty　　　　**Self-Organizing**　　　　**Low Certainty**

As the level of certainty increases, the degree of control and direction increases. As the level of certainty decreases, things become unknown, unclear, and disconnected. It is in the middle of the continuum where there is both some degree of order and disorder where strong relationships form based on free choice, where learning and growth occurs, and where mutual influencing takes place.

These are the three Certainty Patterns – High Certainty, Self-Organizing, and Low Certainty. As a continuum, there are patterns along the continuum, some with more or less degree of certainty.

What moves things along the continuum are changes in the three conditions of self-organizing – Containers, Differences, and Exchanges. For example, imagine three classroom situations:

High Certainty – Desks are in a row, the curriculum is tightly planned and the students have no choices on how they will learn the material presented to them.

Low Certainty – There is open space; students are given materials to study and experiment with but with no direction or help.

Self-Organizing – as in a Montessori school, students have freedom to chose projects and are given support as needed. The creative interactions among students and between the instructors and students are key.

Certainty Patterns in Baking Bread

To reach a state of self-organizing the preparation of the dough needs sufficient planning and organization to insure consistency. It is important to mix the ingredients in the right proportions and in order. If the mixing and kneading of the dough is rushed and not given enough time for the ingredients to integrate, the bread will be lumpy. These are examples of exercising a high degree of order (high certainty).

The dough also needs enough time, space, and freedom to develop, but if left too long and without kneading the "crumb" (technical term for the inside of the bread) will not have enough structure or air to achieve the texture we like. There may be huge air pockets resulting in many big holes in the crumb. These are examples of a lack of order (low certainty).

A successful baking process will let the yeast do its work of leavening the bread and baking will have just the right amount of time – not too long or too short. These are examples of self-organizing – not too much or too little order, structure, and control.

Summary of Certainty Patterns

At times, we are in situations where we are very certain and where there is a lot of order, predictability, or organization. We and others are in control of the situation. These situations can give us comfort and security but they can also become stifling.

At the other extreme, we are sometimes in situations where we are very uncertain and things are very unorganized and unpredictable. No one is in control. This can give us complete freedom of thought and movement, but it can also make us very insecure.

In between these extremes is the situation of self-organizing where we can learn, make relationships, and adapt with an acceptable degree of uncertainty. We can engage the sacred in all three certainty patterns but in the self-organizing space we are more likely to make new connections within ourselves and with others that will be life-enhancing.

Stretch and Fold

We have seen how self-organizing is affected by the three conditions of container, difference, and exchange. We have also seen how the patterns of certainty and uncertainty emerge as one or more of the three conditions are affected. We now explore how stretch and fold is a strong metaphor for affecting the three conditions. Think of an umbrella (**Figure 5-8**) as it stretches and folds.

Figure 5-8 Umbrella

We stretch when we need to learn or want to identify new possibilities and fold back when we need to integrate the learning and incorporate the new possibilities into our daily lives. We stretch to see the future and fold to bring the insights to the present. We stretch to focus on differences and fold back to see if we can make use of the insight. We stretch to pursue our dreams and fold back when we need to pay attention to current reality.

A self-organizing process stretches and folds to bring together any points that are far apart or, conversely, to separate the points. For example, when a conference breaks up into small groups, it is stretching. When the small groups come together and the comments from the small groups are integrated, it is folding. When someone summarizes the conference proceedings, it is folding even further. In terms of the certainty continuum, we stretch into the unknown to break free of the controls that have kept us locked in, and we fold back toward the certainty of a "new" order after we have established new patterns of learning and growth.

Stretch and Fold in Science

Scientists and inventors have moments when they stretch into "Eureka" moments of clarity and insight and fold back into the mundane. Einstein had deep meditative "thought experiments" and Darwin hesitated for decades about publishing his insights from the voyage of the *Beagle*. Long periods of deep thinking stretched their categories and boundaries and then they folded to practical theories like $E=mc^2$ and evolution. Scientists need to be flexible enough to comprehend something beyond what they know.

When there are conflicts in thought, even if two points of view are far apart, a process of folding and stretching will bring the two points close together. This is achieved by creating an interleaving of the two positions with a fractal structure in such a way that there are more pathways between any two points under consideration.

Stretch and Fold in Baking Bread

As a baker stretches and folds the dough, the differences are distributed and integrated. If the dough feels loose or weak, the baker will give it another stretch and fold before putting it back in the mixing bowl. The whole purpose of stretch and fold is to strengthen the dough. This is done over time. The number of stretch and folds depends on the dough.

In mathematics, this process is called the Baker's transformation -- similar to what a baker puts dough through. Imagine rolling the dough out so that it is thinner and longer and then cutting the elongated design into two pieces, placing one atop the other and repeating the process. Mathematically, the result from stretching and folding is chaos. In this analogy, the dough corresponds to a set of initial conditions that are sufficiently near to one another that they appear to be continuous. If we impose a pattern on the dough, by repeating the process 20 times in which the pattern is stretched and folded on itself, the pattern will be spread over 1,000,000 layers of dough. This yields a structure with an infinite number of layers to be traversed by the various pathways. The structure becomes very rich as the pathways diverge and follow increasingly different paths.

If we apply this metaphor to interacting with the sacred, any pattern of thought we have about the sacred, if stretched and folded, will give us a much richer view of the sacred in our lives.

Summary of Stretch and Fold

Stretching and folding our learning, putting one idea on top of the other as in Baker's Transformation, will yield a rich interlacing of ideas and experiences.

Introduction to Chapter 6 Religious Examples

This chapter applies the three complexity methods to a deeper understanding of examples drawn from the religious literature and experience.

Chapter 6 Religious Examples

In this chapter we will use religious examples to demonstrate the relevance of the complexity methods for increasing sacred awareness. The first example about the Triune God, three persons in one Supreme Being, illustrates the value added by all three complexity methods. Each method is then applied to two additional examples.

Although the examples are drawn from Judeo-Christian texts and concepts, the themes have general applicability.

Triune God

The story: [Excerpted from Gresham Machen (1881-1937), *The Christian faith in the modern world* (1978). Machen was Professor of New Testament, Princeton Theological Seminary and Westminster Theological Seminary].

"Christianity teaches that the Father is God and the Son is God and the Holy Spirit is God, and that these three are not three aspects of the same person but three persons standing in a truly personal relationship to one another (**Figure 6-1**). There we have the great doctrine of the three persons but one God. That doctrine is a mystery. No human mind can fathom it. Yet what a blessed mystery it is! The Christian's heart melts within him in gratitude and joy when he thinks of the divine love and condescension that has thus lifted the veil and allowed us sinful creatures a look into the very depths of the being of God."

Figure 6-1 Symbol of the Trinity

A **container** is closely associated with the function of God as the creator of all of the containers in the universe. Significant **differences** can be linked to Jesus who breaks down barriers, crosses boundaries, creates new possibilities for discipleship, and transforms the Law with a gospel of love. The **exchange** condition can be seen in the functions associated with the Holy Spirit. The Spirit fosters exchanges through building relationships, through our sharing and caring.

If we retain a rigid notion of the Trinity (**high certainty**) that prescribes God's commands, what Jesus would do, or where the Spirit is leading us, we may miss the perichoresis or "dance" of the Trinity which creates new metaphors and faith language. Re-imaging the faith and questioning assumptions is activity in the **self-organizing** space. If we are unconnected and separated from the Trinity (**low certainty**), or emphasize one aspect to the neglect of the others, we miss the wholeness produced by engaging with all three aspects of the Trinity in the **self-organizing** space.

If we **stretch** to see that the Triune God is not a static object but an event, an interaction of the Father, Son, and the Holy Spirit, we can more clearly see the functions in the Godhead. The Father is the force and energy that is continually creating the cosmos; the Son breaks down barriers and opens new possibilities; the Holy Spirit raises our consciousness and awareness and connects us in new relationships. We **fold** to experience the sacred unity of these forces that creates wholeness, which allows us to re-image our faith.

Value added to Understanding the Trinity

The complexity methods provide insight into the dynamic nature of the Trinity. The linkage of the three conditions of self-organizing is similar to what is usually ascribed to the three persons of the Trinity. The certainty continuum suggests that the doctrine is best experienced in the self-organizing space where we are free to emphasize each aspect of the Trinity as needed. At times, we need to stretch to value God as the source of the awesome mystery of creation. In other moments, we stretch to break down barriers as Jesus taught. In other instances, we stretch with the Spirit to connect with those around us in community. We fold into all three persons as we reconstruct our sense of the sacred. If we reflect on each aspect of the Trinity, we are reminded of the guidance we receive from each aspect, and our sense of the sacred is enhanced.

Insights about the Sacred

The scope of the Trinity helps us appreciate the awesome breadth of the sacred. The creativity in the universe is seen in God's sustaining of creation, in Christ's inclusiveness and sacrificial love, and in the impulse from the Spirit to connect with everything and everyone.

Conditions of Self-organizing in a Blessing

To illustrate how the conditions of self-organizing are part and parcel of our current experience of religion I will cite a blessing by Macnab (2004):

> May the God you see in all the colors of creation arouse in you a sense of awe and wonder.
> May the God who is a sacred presence be real to you.
> May the God who is a source of inspiration and courage keep calling you forward.
> May your God go with you, and bless you.

In this blessing the **containers** are God in all the colors of creation and God as a sacred presence. The significant **difference** is "source of inspiration and courage". The transforming **exchange** is "arouse a sense of awe and wonder, be real to you, keep calling you forward, and go with you and bless you."

Note that exchange is the dominant aspect of this blessing. The purpose of the blessing is to provide assurance, to energize, and foster a continuing relationship with the sacred. To be effective all three conditions of self-organizing must be there. Any sermon, liturgy, or song could be analyzed to determine whether the conditions of self-organizing are present and which conditions are emphasized. The purpose of this analysis would be to deepen our engagement with the sacred.

Conditions of Self-organizing and Divine Presence

A friend described her experiences during the civil rights march on Washington in 1968. As one of the few whites among many black Americans, she experienced a profound peace and love. As she tearfully recounted what happened I was struck by the sense that she experienced a divine presence. I began to think about what happened that created a powerful inner experience for my friend and for others in the civil rights movement.

What happened that made my friend's experience so moving? The March on Washington was a sanctioned protest that had a focus and purpose (**container**). The mission of the March was a cause that had captured the hearts and minds of many Americans (**difference**). The words, support, and feelings of mutual love were transformative (**exchange**). The persons in the March created change on many levels – in individual hearts (**Figure 6-2**), in the policies of organizations, and in the laws of the society.

Figure 6-2 Together

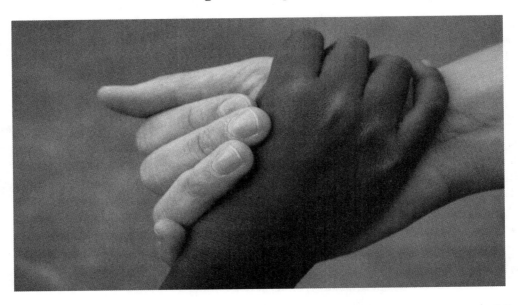

So why do I think about divine presence during the March? An important aspect of the divine in this instance was its teleology, which was defined by its spirit. A crowd gathered with a purpose **(the container)** and the quest for radical change **(difference)** could have so easily (and sometimes did) led to a destructive riot (e.g. Chicago in 1968 or Haight-Ashbury). The **exchange**, however, was expressing a particular kind of spirit: one of unity, justice, equality, etc., which leaders like Martin Luther King insisted was precisely the spirit of the Triune God. A negative spirit, in contrast, would result in a far different exchange. A negative spirit of coercion, greed, denigration, and rejection would have resulted in a very different experience for my friend.

Our actions, individually and collectively, create the necessary conditions for our awareness of the sacred. As was demonstrated by the March, if we create containers that are inclusive of others, pursue significant differences like social justice, and positively engage with others, our awareness of the sacred is enhanced and personal and social transformation is more likely to occur.

Certainty Continuum in Forgiveness

The Story: Joseph's brothers envied their father's favoritism, so they sold Joseph into slavery and tried to kill him. But within a few years the brothers were reduced to beggars while Joseph had become Pharaoh's right hand in Egypt. When their fratricide was exposed, the brothers fully expected that Joseph would punish them. But Joseph forgave them. Joseph believed that God had a larger purpose for good beyond the private wrongs he had suffered: "Don't be afraid. Am I in the place of God? You intended to harm me, but God intended it for good" (Genesis 50:20). At least four times he reassures his nervous brothers, "it was not you who sent me to Egypt, but God" (Genesis 45:5, 7, 8, 9). The story concludes: "Joseph reassured them and spoke kindly to them."

The creative nature of God is to forgive and respond with love. Joseph had many options and plenty of time to consider how to deal with his brothers. He knew the wrong they had done but he also knew what God had done for him.

Joseph was living open to God's presence and living within the boundaries of his faith. The encounter with his brothers could have pushed him into a tight, **high certainty** space where he controlled all of the options about how he could deal with his brothers. He could take revenge, humiliate them, or even enslave them. But when his brothers asked for forgiveness Joseph saw that God had used his brothers' action for Joseph's own good and for the good of many others. Joseph could have chosen life or death. He chose life and lived God's purpose into a **self-organizing** reconciliation space.

Certainty Continuum and Faith

Regarding the sacred, when we are in a state of **high certainty**:

- There is no mystery

- Our sense of God is constrained

- We are in firm control of our beliefs

- We relate to the sacred through comfortable religious practices

When we are in a state of **low certainty**, everything is mysterious, unexplainable, we feel unconnected and unconstrained.

In a state of **self-organizing** our awareness is unfolding, our beliefs are flexible, we create new patterns of relationships, and we are in a state of learning.

Being anywhere on the continuum can be beneficial. There is no right or wrong. At times we need to feel grounded, safe and confident. At other times, we can take risks, open up our boundaries and explore without constraints. But if we stay in either end of the continuum we will miss interacting with the sacred in creating new patterns that are life-enhancing.

For example, if our goal is to have a dialogue with persons who may not share our views of the sacred, we need to move from the high certainty space and suspend our dogmas and doctrines, feelings of superiority, and clarity of our positions, and at least temporarily move in to the middle space of self-organizing. This middle space can be messy but it is the space in which we can share our religious experiences and creative ideas, as well as show our interest in including and learning from the experience of others. After the dialogue is over and misperceptions cleared away, we can integrate the new perspectives into our core values.

If we start from the other end of the continuum of low certainty, in order to have a productive dialogue we need to identify some topic that will focus the discussion or at least wait for a pattern to emerge with which we can engage. The interfaith dialogue recounted in *The Faith Club* by three Christian, Muslim, and Jewish women (Idilby, Oliver & Warner, 2007) is a good example of this experience.

Stretch and Fold and the Kingdom of God

The story: [Excerpted from Marcus Borg, 2003]. "The kingdom of God is a God of love and justice whose passion is for our life together as demonstrated by the message and passion of Jesus. The kingdom of God is not about heaven; it is for the earth, e.g. in the Lord's Prayer – "Thy kingdom come, they will be done, on **earth**, as it already is in heaven. It is what life would be like on earth if God were king and the rulers of this world were not; God's justice in contrast to the systemic injustice of the kingdoms and domination systems of this world. **(Figure 6-3)**. The kingdom can be seen in the power of God active in Jesus' work, the presence of God, a community."

Figure 6-3 Justice

The kingdom is God's creativity working to transform us into beings of higher consciousness, able to articulate and demonstrate God's presence on earth. Demonstrating love and caring in relationships, e.g. caring for those at the bottom of society, will extend the kingdom. We want to participate in the kingdom, but instead we often fill that spiritual hunger with consumption and distractions.

We can speculate about what life would be like on earth if God were king. What would the social and economic systems be like? If the kingdom is here on earth, we have the challenge to **stretch** into creative involvement in the kingdom. We can stretch to affect the conditions of self-organizing that extend the kingdom. As God's agents on the planet, what are the actions we are called to do to participate in the kingdom? As we answer this question, we **fold** to living our life with greater commitment to do something that serves the greater whole.

Stretch and Fold in Sacred Unity

As we interact with the sacred we stretch to learn more about ourselves, discover new meanings, develop greater clarity about the direction we are headed, are able to tell our story to others, gain a greater perspective about our place in the universe, and push into significant differences.

We fold in this learning and perspective by holding on to the new, finding common ground with others, and developing conclusions about what is right for us. Then we pause and bring all these factors into clear proximity, finding new connections and discovering new possibilities.

When, in Galatians 3:28, St. Paul said, "Because all of you are one in the Messiah Jesus, a person is no longer a Jew or a Greek, a slave or a free person, a male or a female", he was stretching into new possibilities **(Figure 6-4)**. By stretching into being like Christ, Paul is saying we fold into equality with each other.

Figure 6-4 Bernini Statue of St. Paul in Rome

63

To help us stretch into the sacred as we engage with others and when we are reading religious literature we can say:

- Tell me more
- What does it mean?
- Give me some examples
- How would that happen?

We can fold and integrate with comments and questions like:

- Let me think about that
- Where's the common ground?
- Can you summarize that?
- How would you apply this?
- What does this invite you to see?

Conclusion

The complexity methods can loosen our boundaries to increase our flexibility, adaptability, and ability to forgive when we are confronted with different views. The complexity methods increase our understanding of the scope of the sacred and the possibilities that are open to us. They can also increase our motivation and impulse to take action.

PART III Application of Sacred Awareness Principles

Sacred Awareness Principles

Figure III-1 is a graph of the five sacred awareness principles that integrate the themes about the "bathwater" in science and theology discussed in **Chapter 3** and the insights from the complexity and emergence concepts and methods discussed in **Chapters 4 and 5** and applied in **Chapter 6**.

We will use these principles in **Part III** to explore developing sacred awareness.

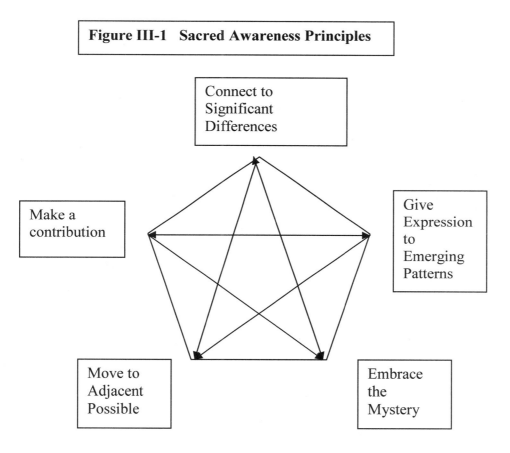

Figure III-1   Sacred Awareness Principles

Connect to Significant Differences

Make a contribution

Give Expression to Emerging Patterns

Move to Adjacent Possible

Embrace the Mystery

The pentagon formed by the five principles shows that each of the principles is connected to the other four. There is no linear sequence. In increasing sacred awareness a person can start with any principle and see where it leads. Below are some highlights of each principle. In the following chapters the principles will be applied to our individual journeys **(Chapter 7)**, in valuing and relating to others who have different belief systems **(Chapter 8)**, in being stewards of the planet **(Chapter 9)**, and in providing leadership in our groups and organizations **(Chapter 10)**.

Connect to Significant Differences (See what is in the bathwater)

Raising sacred awareness inevitably requires an interest and some courage in encountering difference, the new, and the unknown. To stretch, evolve, and develop our competencies in engaging the sacred, we must be willing to "see what is in the bathwater" and be willing to let significant differences change or reinterpret aspects of the "baby" in the water. The new thinking in both theology and science can open our eyes to what has been unknown or hidden from us.

Give Expression to Emerging Patterns (Prepare the bath)

The interactions of agents, including people, things, and ideas are continually creating new patterns all around us. Noticing and participating in the emergence of some of these patterns and giving expression to them is what it means to be an agent in the creativity in the universe. This is "preparing the bath".

Embrace the Mystery (Create waves in the bathwater)

Embracing the mystery is not about embracing the unknown. The mystery is the awesome self-organizing processes in the universe that creates what we hold to be sacred. These processes are surprising and uncertain (they "create waves"), yet they are key to providing us with the insights, emotions, and awe we associate with the sacred.

Move to the Adjacent Possible (Get a big enough tub)

Developing sacred awareness requires attention to our containers. Do our images of the sacred limit what is possible for us? Are we empowered by the boundaries we have set for the sacred? Is our "tub" big enough or is it too big? Are the stories we create inclusive and relevant for our times? Acting on our insights and awareness requires us to move into the adjacent spaces that are possible for us.

Make a Contribution (Be the baby – splash and play)

For us to co-create the sacred "baby", we need to use our unique talents and connect with others in community and across the boundaries that separate us. These exchanges can be transformative. The synergy that is possible makes the whole much greater than the sum of the parts.

Chapter 7 Personal Sense of the Sacred

The purpose of this chapter is to help individual readers to clarify their own meanings of the sacred. Readers can explore their own sacred awareness, incorporating and integrating their own important values and beliefs with an understanding of cosmology, spirituality, emergence, consciousness, and the sciences of complexity.

I will use the five principles of sacred awareness to guide the discussion.

Connect to the Significant Difference: See What is in the Bathwater

One metaphor for God is that forces in nature create large effects from small factors **(Initial Conditions)**. This is useful for discerning what is present in the bathwater that may make a big difference for ourselves and others.

Before we can talk about patterns, we need to know what we are looking at. Bubbles make different patterns than strings. Western psyches make different patterns than Eastern. Languages that use alphabets (English, Greek) create different ways of expressing thoughts than those that use pictographs (Chinese). Seeing the bathwater means seeing the particulars of our lives as being a complex amalgamation of unique and discreet choices, events, tastes, physical properties, cultural conditionings, etc. that are unique for each individual.

Once we are aware of these discrete elements, the patterns we have created using these elements can be seen less as a matter of predetermined fate than a matter of creative composition. We can slow down and see our peculiar constellations of realities and events and let go of judgments, orthodoxies, and dogmas about what <u>should</u> be **(Figure 7-1).**

Figure 7-1 Making Choices

We can identify and amplify significant differences and establish priorities among the options and choices we face daily. If God and the sacred can be seen in the effects that come from small events or surprises, there is hope that we do not have to answer life's questions alone.

Give Expression to Emerging Patterns: Prepare the Bath

Senge (2005) quotes Brian Arthur, an economist, as saying that once people arrive at a place of genuine knowing about who they are **(Figure 7-2),** they inevitably operate out of a larger intention which is simply doing what obviously needs to be done. To develop a greater sense of the sacred we need to give voice to the new patterns we see emerge. As we let go of certainty we can experience the sacred in new patterns. The distance between us and the sacred can be reduced.

Figure 7-2 The Wonder of who we are

Morwood says we need to inquire less about who God is and shift attention to the wonder of who we are. Morwood says we are a life-form giving "God" a way of coming to expression. Our reflections, our times of prayer, our times of community worship are how we prepare ourselves to move to where the sacred is most evident -- within oneself. For example, when we identify and honor important symbols such as a sacred holiday we can rekindle the vitality and purposefulness of life. When we take time to pray and meditate we can shed the busyness that crowds our life.

"Thin places" (Borg, 2003) are where the two layers of reality, the world of our ordinary experience and the sacred, meet or intersect.

To be fully attuned to the sacred probably takes the preparation described above: deep meditation, discipline (e.g. 40 days in the wilderness), and extraordinary dedication. Christ and some of the saints point to what this may mean. But we do not need extraordinary effort to see the astonishing mystery embedded in life. Most of us can sense the sacred in the "thin places". They are the places "where the boundary between the two levels becomes very soft, porous, permeable. The veil momentarily lifts, and we behold God, experience the one in whom we live, all around us and within us." (Borg, 2003). The thin place may be in certain geographical locations or in nature, music, poetry, dance, or worship. It could be a mountain top experience, a new relationship, in your garden, seeing the sky at night, witnessing the birth of a child, or beholding a galaxy revealed by the Hubble telescope. When we personally become thin (vulnerable and broken from sickness, loss of a loved one, poor choices, or just plain living) we find the sacred.

Embrace the Mystery: Create Waves in the Bathwater

The turbulence in the bathwater moves things far from equilibrium toward a new pattern. Small causes that have big effects are the significant differences that start the process of self-organizing. When the conditions are right we experience the effects of the sacred. If we are still, open, and attentive we can be aware of the differences in the bathwater.

Carl Jung believed that the psyche is a self-organizing entity stirring to continuously unfold an innate destiny (Conforti, 2003). As the psyche self-organizes from repeated exposure to stimuli, consciousness self-organizes and emerges. What is it that organizes the emergence? Jung spoke of the numinous moment in time that something has happened – perhaps a tipping point, epiphany, or personal revelation that triggers emotion leading to new patterns of self-organization.

Borg (2003) quotes Thomas Merton, a 20th Century Trappist monk:

> Life is this simple. We are living in a world that is absolutely transparent, and God is shining through it all the time. This is not just a fable or a nice story. It is true. If we abandon ourselves to God and forget ourselves, we see it sometimes, and we see it maybe frequently. God shows himself everywhere, in everything – in people and in things and in nature and in events. It becomes very obvious that God is everywhere and in everything and we cannot be without Him. It's impossible. The only thing is that we don't see it.

Merton is saying that the sacred is all around us but we often do not see it. The "Eureka" discoveries of scientists or creative insights and the intuitions we have in dreams or while taking a shower are times when the sacred breaks through (**Figure 7-3**). Carl Jung (1989) would have identified thin places and the sacred with the "numinous", something wholly other that produces an intense feeling that there is something that cannot be seen.

Figure 7-3 Light Shining Through

Clayton (2004) and Jeeves (2004) believe that God's presence in the ongoing creation is seen at the level of the individual person. Every individual unavoidably encounters the divine every instant. That is the meaning of "Divine"—the one event every other event through all time encounters.

Move to the Adjacent Possible: Get a Big Enough Tub

Transformation emerges when a person stretches to a new set of adjacent opportunities in which something different is chosen and then folded back to our daily life. When we interact with the sacred, the challenge is to recognize the opportunities that are in front of us and stretch to them. This requires some risk and a learning curve that many of us are not willing to take on. Sticking with the status quo and taking the world on "faith" is easier to do. Our faith and courage to act are, in fact, sacred as we stretch to live life forward, even though, as Kauffman (2008) says, we cannot know everything we need to know. We live our lives in faith, a committed courage to get on with life anyway. Plato says we seek the good, the true, and the beautiful. Our faith and courage are, in fact, sacred – they are our persistent choice for life itself. We do, in fact, stretch forward into mystery and fold back with assurance that we are part of the sacred.

As life-forms that give the creativity in the universe (or God) a way of coming to expression (Morwood, 2007), we can jump to thin places where we have contact with the sacred. Those thin places are different for each person. Identifying symbols and images that are meaningful may suggest a direction of where to jump. Jumping to the adjacent possible is a stretch towards mystery that takes faith and courage. Kauffman calls this faith and courage -- sacred, our persistent choice for life.

We move to the adjacent possible in the search for purpose and meaning as we evolve and adapt, as we establish our daily goals and expectations, and as we create what sustains us and let go of what keeps us from evolving.

Make a Contribution: Be the Baby –Splash and Play!

A sense of the sacred in human life can help to create what Geering (in Jones, 2005) calls a "meaning system" to fill a spiritual void as we are impacted by the new cosmology. A better understanding of the sacred can help answer such life questions as "who are we, what is our purpose?" Geering says that the attempt to live only on "the accumulated spiritual capital of the past" has given rise to fundamentalism in all religions. The fear of change sometimes leads to implausible beliefs and destructive practices. Geering says:

> Though humans have had little to say about developments over the four billion years of life's evolution, the explosion of knowledge and technological power in the last 100 and 50 years has made us responsible for how the story of earth will unfold from here on. The story of earth depends on the decisions and actions of humankind. Evolution is in our hands.

Geering is suggesting that we can create the initial conditions that will let us co-create with the sacred in our continued evolution. The sacred is a vital part of our struggle with the questions of purpose or meaning in human life. For many persons wandering the halls of institutional religions, the dogma about the sacred is illogical and unpersuasive, given what we have discussed. Those who have rejected religious paths often end up in despair about the inhumanity of man. Many are soothing their despair and anxiety through material acquisitions, drugs, exotic vacations, or a focus on their own or others' achievements, e.g. Olympics (Young-Eisendrath & Miller, 2000).

Conclusions

We are very likely to experience the sacred in our personal lives when we:

- Search for our purpose and meaning.
- Establish daily goals and expectations.
- Create what sustains and let go of what keeps us from evolving.
- Establish priorities among many daily options and choices.
- Experience forgiving and being forgiven by others.
- Experience the turbulence in the environment that moves things far from equilibrium toward a new pattern.
- Seek and receive transforming information.

- Remove barriers to interactions and the exchange of energy with others.
- Receive and give comfort and joy in our relationships.

We develop our vision of the sacred during our life journey. The "baby" emerges from the bathwater. The images, words, and theories we use along the way create the conditions for the emergence of our vision.

Figure 7-4 is a symbolic image of a sacred moment. A rainbow at the end of a bank of solar panels is symbolic of the self-generating, renewing creativity in the universe.

Figure 7-4 Creativity in the Universe

Introduction to Chapter 8 Interacting with Others about the Sacred

This chapter discusses how a greater sense of the sacred can help us work across the barriers that separate us from one another.

Chapter 8 Interacting with Others about the Sacred

In this chapter we apply the sacred awareness principles to how we view those who are different from us. Our sense of the sacred is affected by the stereotypes we hold, the interactions we have (or do not have), the moral choices we make, and how we deal with social justice issues.

Connect to the Significant Difference

Deepening our appreciation of difference requires us to stretch our skills in contemplative practice. Contemplative practices that quiet our senses, such as silent prayer, meditation, and praying the scriptures, help us to suspend evaluative thought and develop compassion for self and others. Reflection promotes humility and emotional balance as core values to open ourselves to the mystery that is within us.

When we encounter difference, we can set aside quiet time for personal reflection, journaling, or intentional pauses in the conversation to provide opportunities for individuals to reflect and clarify their own points of view and give differences an opportunity to emerge.

Give Expression to Emerging Patterns

Emergence has profound implications for communicating about the sacred. Instead of trying to change individual beliefs about the sacred, we can create the conditions for emergence. Some of the things we can do to influence these conditions are:

Containers

- Establish over-arching objectives; reinforce shared values and visions. Articulate the common ground that holds the group together.

- Introduce concepts and activities that provide a unique context for the interaction.

- Recognize the potential for turbulence that can come from engagement with difference.

- Have no agenda – be willing to respect and accept what others say, along with their limitations.

Differences

- Deconstruct differences to reveal the meaningful distinctions between the categories. For example, a person might be a Christian or a Muslim, but both are interested in following the tenets of their religion although they take their inspiration from different sources.

- Do not let a single difference predominate. A small number of significant differences help focus the conversation, but when a single difference predominates, the conversation tends to collapse into an argument.

- Focus on differences that can be negotiated if people become stalled over a significant difference, preventing new patterns from emerging

- Frame differences so people can see that within differences there are other important differences, for example, the differences that can exist within a religious denomination.

- Identify the systemic policies and procedures that separate faith communities.

- Recognize that each person has a unique life experience and may not be referring to the same thing, even if the words are similar.

Exchange

- Establish cross-faith learning groups to focus on specific differences and similarities of shared, systemic interest such as *The Faith Club* (Idliby, Oliver & Warner, 2007).

- Provide mentoring and training programs that are readily available to all those who wish to discuss questions about their faith.

- Say clearly what you mean rather than "beat around the bush." .

- Actively listen -- give your whole attention to another person, the subject matter, and the shared experiences as if nothing else matters.

- Speak from the heart and share your authentic self.

- Don't say what you think others want to hear.

- Silence can be an inauthentic exchange, e.g. not confronting bigotry.

Embrace the Mystery

In interacting with others about the sacred it is useful to determine what others are certain about and where there is room for discussion.

When we decide whether to take on a moral or religious issue, and which battle to fight, e.g. issues of homophobia, racism, sexism, or religious bias, etc., moving to a self-organizing space is essential. If we stay in a high certainty space, dogmas trap us in a quest for perfection. Moving to the very uncertain and unconstrained space, however, avoids the dilemmas inherent in these issues. A self-organizing position would ask these questions:

- Am I acknowledging both sides of the issue?

- Am I keeping an open mind?

- Can I accept that there are no right answers and the outcome is not pre-ordained?

- Is this a battle worth fighting? How significant is this issue?

- How will my decision impact others?

- Can I live with the ambiguity until I know what to do?

- Have I talked about this dilemma with others to make a collective moral choice?

- What have I done to be aware of the sacred in my struggle with this?

- Am I able to state an opponent's point of view in language acceptable to my opponent?

Move to the Adjacent Possible

As we interact with those that are different from us, are our boundaries and stereotypes too rigid? Do we think of them as the "bad guys"? **(Figure 8-1)** Do we see that religious ideologies, race, economic or other conditions result in violent responses? Can we identify the processes that perpetuate fear and rejection of others and make new choices that will generate different responses?

Figure 8-1 A Bad Guy??

Are we open to respecting the religion of others? Are we imposing our culture? What can we do about the fear and hate in ourselves and in others? Do we avoid having any contact with some groups? Are the contacts we have one-sided without listening to other views? Are we open to dialogue and mediation to understand the issues? Can we choose alternative ways of obtaining information?

Fundamentalism in Christianity, Islam, and Judaism, as well as among the religions of the East, continues to be destructive. We need to break out of the boundaries and religious certainties of the past and embrace a new consciousness about the sacredness of the "other". We can be more alert for situations where others are excluded and stand up for inclusion. We can change the conditions that create social injustice.

Make a Contribution

Dialogue is an essential skill for interacting with persons who have different views of the sacred. For example we can avoid conversation stoppers like "It's God's will…". Instead we can use questions like "I'm wondering…" to open a conversation. To increase our sacred intelligence we can critically examine the assumptions that underlie our own faith and the faith of others. In dialogue we can bring the understanding from one faith to bear on another. The questions, opinions, and perceptions of others help us think through our assumptions and reasoning.

We can start with any pressing question (**Figure 8-2**). We could try to understand the relation of Christ to Jews or Muslims. We could explore the question why good people and sincere believers suffer. We could discuss the efficacy of prayer. We could start with views of an afterlife. It is less important where we start than having a good process for our thinking and discussions.

Figure 8-2 Start with any Question

Here are some guidelines for a session devoted to dialogue with persons from other faiths:

- **Speak concretely**. If we tell stories that provide detail about our experiences with the sacred, the listener can tune into the deeper feelings and assumptions underlying what is being said

- **Listen actively**. As "the other" tells his/her story, listen with a spirit of openness and active attention to the words, the affect, and the body language.

- **Suspend judgment**. As each person takes their turn, suspend your judgments and hold back from expressing them if they arise. Judgments can be expressed in follow-up discussions that have been informed by the expressions of the rich information from the dialogue session.

- **Acknowledge or express assumptions**. When you speak, make your assumptions as explicit as possible. By bringing the deep assumptions into the open, the diversity of opinions and beliefs in the group can be revealed so that a true consensus and shared learning can emerge.

- **Engage in helpful inquiry**. Make honest inquiries to understand more deeply where the speaker is coming from. In follow-up discussions the group can assist speakers in uncovering assumptions of which they may be unaware.

- **Be open to outcomes**. Let conversation flow in whatever direction is necessary to open new ways of seeing the sacred, as long as the overarching objective and atmosphere of respect are honored.

- **Slow down and reflect**. Slow the pace of the conversation so that attentive listening, inquiry, and reflection can occur. Using the silence between the stories helps all members to open their minds to new possibilities.

Mapping the Sacred

Individually and collectively, each person's view of the sacred is influenced by three factors: worldview, underlying rules, and perceptions of reality. For example, the writers of the gospels of the New Testament were influenced by the prevailing **worldviews** of the Jewish community. They were influenced by the perception of **reality** that the earth was flat and part of a three-story universe. The commonly accepted underlying **rules** were the Ten Commandments and the laws and practices of the Jewish nation.

Worldview: Our foundational perspective, how we think things are supposed to be, and what we think is true and useful for us. A worldview also provides the filters that influence our thinking. These filters include our genes, family, education, culture, history, profession, personal experience, values, self-concept, and religion. For example, our worldview shifts when there is a terrorist shooting. We thought that public buildings are safe places to be; now we have metal detectors. Decisions about where to take a vacation are affected by this worldview.

Regarding the sacred, examples of various worldviews would be –

- Science is the only reliable way to truth.
- There is a providential presence in the universe
- There is a sense of mystery in the universe that is exhilarating
- Evolution does not preclude a deity nor does it require one.
- The universe has a purpose.

Underlying Rules: These are the rules (social norms, formal, informal, and regulatory) we believe we should follow. These could be laws, policies, procedures, agreements, ethical principles and cultural expectations. For example, when driving a car I am aware of the speed limit, the rules about passing other cars, the safe distance between cars, and whether I can turn right on red. I keep these rules in mind but I also consider the reality of the situation (wet slippery pavement), and my own world view (I am a safe driver).

Regarding the sacred, examples of various sets of rules would be –

- The Bible is infallible, inerrant and directly inspired by God
- Take the Bible seriously, not literally
- Practices and customs of a faith tradition

Perceptions of Reality: These are what we see, what we understand about the world out there, objective data, observations, pressures that we experience with our five senses. Parables, stories, and metaphors can help us imaginatively re-experience reality by exposing us to an unimagined alternative view of life.

Regarding the sacred, examples of various perceptions of reality would be –

- Reality is purely mechanistic and predictable; with sufficient data all outcomes can be predicted.

- Reality is unpredictable and hostile; forces affecting life can be manipulated by appealing to supernatural powers.

- Reality is both physically logical and experientially mysterious. Both scientific discoveries and symbolic and metaphoric language are needed to understand deeper levels of reality.

- Reality is an unpredictable amalgam of inter-connected entities that create continuously changing but inherently self-organizing patterns.

Mapping Your View of the Sacred

To help our exploration we can use a map. The questions in **Figure 8-3** help us to explore our own views of the sacred and the views of others who are different from us. We can use this map to overlay alternative worldviews, assumptions, and perceptions of reality that impact our connection with others about the sacred. For example:

- *Competing worldviews* – "There is a providential presence in the universe that sustains it" versus "There is a sense of mystery in the universe that is exhilarating that doesn't require a deity."

- *Conflicting assumptions* – "We pray to influence the actions of God" versus "We pray for our own greater understanding."

- *Contrasting views of reality* – "Ultimate reality is personal – intelligent, capable of love, and making promises" versus "Ultimate reality is impersonal – patterns emerge from unpredictable interactions."

Figure 8-3 Questions to ask yourself

What are your <u>worldview and filters</u> about the sacred?

- Church/synagogue/mosque
- Profession/education
- Personal experience
- Other

What <u>rules</u> do you follow regarding the sacred?

- Commandments
- Tithing
- Prayer
- Confession
- Keeping Kosher
- Doctrines
- Tradition
- Ethics
- Other

What are your <u>perceptions of the reality</u> of the sacred?

- Scientific understanding
- God's presence in ongoing creation

- View of nature
- Other

Making Decisions

As individuals, when we make a decision about the sacred, we use our worldview, rules, and perceptions of reality.   We usually seek a balance among all three. We may need to shift one or another to develop coherence. The three factors are in a dynamic relationship; a shift in one generates tension that encourages a shift in the others. This process may be conscious or unconscious. What we repeatedly do reinforces the patterns over time.

When a group makes a decision, it mediates the maps of all its members. Over time, a consensus emerges for the group that may be different from the sum of the individual maps. Group maps impact individual maps over time. If groups identify the factors that are influencing the members, together the group can generate a shared understanding to guide their work together.

If we want to influence the decision of a group we can choose to intervene on the one factor most likely to produce change. For example, if we want to change the worldview of a group, changing their perception of reality is likely to be more effective than directly confronting their worldview. A religious worship committee that has a worship service format and schedule that has served the church well for many years is not likely to change when presented with another format and schedule. However, pointing to the declining attendance figures may drive attention to the reality of the situation.

Conclusions

As a society, conflicting worldviews about the sacred have contributed to the many walls of separation among races, religions, social classes, and regions that are jeopardizing and compromising the future of life on earth. We need a sharable worldview about the sacred to reduce the drive toward fundamentalisms as our diverse civilizations collide.

A renewed sense of the sacred will help us work across religious, ethnic, racial, class, gender, sexual orientation and other boundaries to expand our humanity until we can become barrier free (Spong, 2007). As individuals, perhaps we can learn to curb the certainty with which we dehumanize those who are different (Nelson, 2005).

Introduction to Chapter 9 Sustaining the Environment

This chapter applies the new sense of the sacred to understanding how our social, environmental, economic, and institutional web is destroying the environment of the planet and what we can do about it.

Chapter 9 Sustaining the Environment

The issues we are facing in global warming, pollution, extinction of species, destruction of rain forests, etc. are the result of unawareness and disregard of the impact of consumption of fossil fuels and consumption of natural resources on our ecological and human systems. We need to change our thinking about what is sustainable. Recapturing a sense of the sacredness of creation and our responsibility as co-creators will help make this change.

This chapter applies the new sense of the sacred to understanding how the environment is being destroyed and what we can do about it.

Connect to the Significant Difference

The creation story of an emergent universe is new. We do not have cultures or traditions that celebrate it, unlike the creation stories from the Bible, the Qur'an, and other ancient writings. The story of our emergent universe stretches us to take responsibility for the planet and its ecosystem. The story shows how connected we are to all living things. We were not added to the earth after a divine being created it. We came out of the earth, which came out of stardust. The scientists and theologians agree on one point: there is a profound wisdom at work in the universe. Brian Swimme (1999) says, "The earth was once molten rock and now sings opera." All creativity and all consciousness arise from the earth itself **(Figure 9-1).**

Figure 9-1 Hawaii Coastline

When the impact of natural catastrophes is assessed, such as the impact of the hurricane Katrina on New Orleans and Mississippi, we need to look for the differences we ignored in all of the small events, thoughts, ideas, decisions, and behaviors that led up to the ineffectual preparation for and response to the catastrophe. Illness, pollution, global warming, etc. are about the choices we have made individually and corporately. On the other hand, we cannot control everything; a sense of the sacred gives us the courage and peace of mind to live in a turbulent world. We can deepen our awareness about affecting what is in front of us and accept the reality we cannot transform.

We need to stretch to comprehend the implications of this creation story and fold back to a new identity as citizens of the cosmos. Swimme points out that we are kin to every being on the planet in terms of energy and genetics. Stretching to comprehend this and fold it back into our everyday consciousness is a daunting task.

To embrace this new creation story we do not have to reject our traditional stories if they provide comfort and support and connect us to our ancestors. However, we need to examine where these stores misguide our actions and jeopardize the future of life on earth.

Give Expression to Emerging Patterns

Our sense of the sacred is a resource for the recovery and the response to what happens. Beutner (2004) in his analysis of the parable told by Jesus about the unknown arrival time of a burglar (Luke 12:35-40) believes the message is that much of what happens, happens whether or not we have made any great effort to forestall or prevent it from happening. Given that chance is built into the universe, Jesus in the parable is advising watchfulness but in the spirit of relaxation and hope rather than nervousness, vigilance, and despair. Beutner believes that surrendering our fears makes room inside our heads and lives for the arrival of unexpected grace, i.e. presence of the sacred.

If we are aware of the emerging patterns and give them voice, this is the story our grandchildren will tell about how we made a contribution to sustaining the planet:

> In a time of environmental crisis, our grandparents were aware of the peril the earth was in. They saw possibilities for transforming the future. They knew that the earth was not environmentally sustainable on its current course. They believed that human activities were the cause of environmental instability. They began to cherish the earth and actively sought ways to reduce their consumption so that all living creatures could live in a harmonious, mutually self-sustaining ecosystem..

Embrace the Mystery

We can best deal with the environmental crisis by creating the conditions of self-organizing. The serenity prayer could be modified to: God grant me the serenity to accept the things I cannot change; courage to change the **conditions of self-organizing;** and wisdom to know the difference.

Consider that we need to change our conditions and loosen our **containers** by reconsidering any stereotypes we have about "being green" and "tree huggers."

- Admit that some religious ideologies limit our caring for the planet.
- Identify the multiple containers that perpetuate global warming and make new choices that will generate different responses.

- Take personal responsibility: do we assign blame only to Big Oil, Wall Street, etc.?

Be aware of significant **differences** between the old story and the new one:.

- Focus on environmental sustainability,

- Respect the views of scientists who study environmental issues.

- Look beyond our materialistic culture: are we imposing our beliefs about use of natural resources on others?

Be open to challenging **exchanges** by seriously engaging environmentalists:

- Listen to other views and dialogue to understand the issues.

- Refuse to be controlled by popular media: find alternative sources and ways of obtaining information.

Move to the Adjacent Possible

Our sense of the sacred can help us to make moral choices about the environment, especially in the "thin" places. Borg (2003) describes how the prophets were "God intoxicated" — they felt the feelings of God and were filled with the passion of God. They were advocates of an alternative social vision to the prevailing domination system.

By becoming "God intoxicated", we can imagine loving creation as God does, and then dare to act. We do not need to fix everything ourselves – we cannot! But each one of us can take on a moral issue and fight at least one battle, e.g. habitat degradation, ozone depletion, loss of forests, loss of arable land, dead zones at river mouths, the increase of lung and breast cancer and asthma in children from toxic environments, melting glaciers, mass extinction of large animals, etc. Increasing self-organizing would reduce the influence of the domination systems that maintain the status quo. We cannot control all of the factors causing these crises, such as population control. On the other hand, we must not succumb to a sense of being overwhelmed. We can take a self-organizing approach which, as we learned in Chapter 8, would ask these questions:

- Am I acknowledging all sides of the issue?

- Is this a battle worth fighting? How significant is this issue?

- Can I live with the ambiguity until I know what to do?

- Have I talked about this dilemma with others to make a collective moral choice?

- Which of the moral issues are sacred to me?

Make a Contribution

Kauffman (2008) says that the natural creativity in the universe gives us a sense of God in which we can all share. From this natural sense of God, he says we can reinvent the sacred as the stunning reality of creativity in nature. From that new sacred, a global ethic could be invented to orient our lives to connect to an emerging global civilization.

Kauffman believes new connections, networks, and niches can be developed for new goods and services to advance efforts to develop organizations, communities, governments, and societies that are able to advance the long-term welfare of this planet and its inhabitants.

Conclusions

The environment of the planet is in peril. So far our collective response to the crises of sustainability has been to continue the trajectory toward unsustainability, even though there are now millions of groups and organizations talking about and taking some action on all aspects of the crises. For example, the Pachamama Alliance (www.pachamama.org) is an international non-profit effort to demonstrate the connectivity of environmental sustainability, social justice, and spiritual fulfillment to motivate individuals and communities to take action.

The adjacent possible for the industrial north is to radically change our assumptions about the inherent value of perpetual economic growth.. If the industrial north reduced its consumption of natural resources, particularly oil, it would reduce the pressure on the rainforest, the home of native people in Central and South America. If our sense of the sacred includes concern about the well being of the planet, this is the kind of self-sacrifice that is required.

Introduction to Chapter 10 Leadership in Groups and Organizations

This chapter discusses how a new understanding of the sacred can help both formal and informal leaders in managing groups, organizations, and communities in times of uncertainty and unpredictability.

Chapter 10 Leadership in Groups and Organizations

This chapter focuses on applications of a new understanding of the sacred that embraces both religious and scientific concepts for leading and managing unpredictability in teams, organizations and communities. Effective leaders in the 21st Century, including both formal and informal leaders, will need to develop the full range of their capacities, including the spiritual dimension of their lives. Groups, organizations, and networks of organizations (including virtual organizations) can be places for engaging with the sacred.

Connect to Significant Difference

When Lewin and Regine (2001) speak about the importance of "deep work", personally and collectively, they are asking about what are the significant differences in our lives and in our organizations:

> Personally it means developing a caring and connected relationship to yourself in terms of your work, reflecting on what you are doing and why you are doing it – listening to your heart's desire, what you really care about. We need to ask ourselves, "What am I passionate about; what is my passion?" This requires us to awaken our imagination of what might be possible, reclaiming our dreams, listening to the longings in the soul, and seeking those caring, connected relationships that will guide us. We need to believe that we have something to contribute. Every individual purpose is needed for the health of the whole.

Collective deep work requires people working collectively in a team, organization, or community to ask themselves what they are about, what they need, and where the sacred is leading them. This requires reevaluating operating policies, business goals, and whether authentic, connected, caring relationships are being developed. The inquiry needs to include the worldviews, the reality of the situation, and the rules that govern behavior in the organization. Deep work addresses our collective interdependence within the organization, between organizations, and with the sacred.

In the life cycle of a product or an organization, the first curve in an S-curve depicts the typical path of formation, rapid growth, and process improvement (**Figure 10-1**). The second curve in the S symbolizes the re-invented organization that has transformed and **stretched** itself for continued success in a changed world. Organizations that do not stretch to see what is in the bathwater of a changing environment do not survive.

Figure 10-1 The S Curve

In the business world, leaders who sense a decline in productivity, who take note of missed deadlines, and who see the trends in other organizations have a chance to reflect on the realities and make the necessary strategic realignments. There are many uncertainties in this analysis, however, and leaders need to stretch their imaginations, stretch into the uncertainties of what is on the other side of the S-curve and take risks in deciding when and where to invest their resources. After stretching, they must **fold** back to introduce any new market strategies or technology they have envisioned or acquired.

Likewise, the notion of stretching into the S-curve is a useful metaphor for the role of leadership in stretching an organization to engaging the sacred.

Individuals have the responsibility for deciding their purpose, but organization leaders can be exemplars and teachers. Leaders can provide lessons from the literature of their faith, but their main asset to their colleagues is to help them think through their purpose in the organization, in their family, and in society. If this is the focus, leaders will avoid giving others false certainty and telling others what to think about the sacred. Leaders can help organization members to stretch into engaging the sacred, not escape into a zealotry that is misleading and dangerous.

Give Expression to Emerging Patterns

Discussion about the sacred, meditation or prayer, and mental and physical exercise change daily interactions. The outcomes can produce productive patterns and an up tick to organization life; consider, for example, the benefit that regular exercise rituals effect in Japanese companies. Charles Handy (1998) calls such disciplines"Turkish baths for the soul." Handy is a former executive of Shell Oil and professor at the London Business School who says:

> Religion like this is a great aid to self-responsibility. It might even be essential. But it is religion without the creeds and without the hierarchies. It is the religion of doubt and uncertainty, offering one the strength to persevere, to find one's own way in a world that is, inevitably, very different from any world that was known to those who went before.

Handy's view of religion as an aid to self-responsibility and strength to persevere (**Figure 10-2**) suggests that the virtues usually associated with faith communities – hospitality, respect, humility, relationships, and caring – can be practiced in secular organizations.

Figure 10-2 Taking Responsibility

Embrace the Mystery

When organizations respond to crises faced by its members, a sense of the sacred in interpersonal relationships, group dynamics, and in a supportive community can make a difference.

"Why do bad things happen to good people?" is a recurring question. Accidents or deaths of persons close to us can drastically change our lives. Economic fortunes can be wiped out by "acts of God." From a complexity perspective, the indeterminate universe necessarily includes disturbances in all of our lives. These disturbances produce variability that enables the biosphere, economy, and human systems to co-evolve.

The changes in one sphere lead to modification or new functions in another. These changes fit together for a time and then evolve further. For example, the ever evolving technical revolution is constantly changing how we communicate and work.

Faith-based institutions are notoriously resistant to change, however. Religious leaders need to be especially aware of this reality to discern whether the community's outreach processes are likely to produce unintended and negative consequences.

Containers:

Does the group's notion of God limit its response to member or community issues? Has the group developed policies or structures that restrict free expression or movement? Do the worship services exclude persons from the community?

Does the music only appeal to a segment of the group? If so, productive change will be unlikely.

Differences:

On what does the group focus? Are broader societal issues on the radar screen? Are there conflicts in the group over trivial format or personality issues? Is the group ignoring "the dead elephant in the middle of the room" and avoiding important issues? Do the members realize that setbacks and sacrifices can give birth to new creation? If artificial means are used to downplay and ignore differences, productive change is, again, unlikely.

Exchanges:

What kinds of contacts and conversations are there among different subgroups? Is there dialogue between group members and others in the community? Are the contacts meaningful? Does the group realize that in partnering with others, they will be engaging with the sacred? Limiting exchanges to those with whom we are likely to agree will limit the possibility of change.

Fox (2006) says that the Spirit is perfectly capable of working through participating democracy in religious structures; hierarchical and dominating modes of operating can indeed interfere with the work of the Spirit. By influencing any of the three conditions of self-organizing we can reduce power and control by religious hierarchies.

Move to the Adjacent Possible

Lewin and Regine (2001) provide a case study of Babels Paint and Decorating Store, a successful company with five locations and 50 employees in the Greater Boston area. The company began and was successful in the 1950's as Babel's Paint and Wallpaper Store, but the owners recognized there was no security in staying in their comfort zone. They found an adjacent possible in becoming a "Decorating" store.

> *Reality.* The owners realized that not everyone can be a do-it-yourselfer. There was increasing competition from the "Big Box" stores that offered lower prices. Everyone has a limited budget for painting, but besides a good price, the customers needed advice in working out their specific situation.

> *Worldview.* The owners believed that every employee had the potential to be excited about what they do at work. The owners had a strong sense of responsibility to the community and wanted to be the best in their profession.

> *Rules.* The policies they developed to actualize their dream gave employees a strong sense of ownership and opportunity to cultivate their passion for customer service. The practice was to help customers discover their inner design professional. An overall rule was to add beauty to people's lives.

The company leaders were aware that a new business climate was emerging. They saw cues about both the need to change and how to do it. The adjacent opportunities that became available were limited only by the imagination of those spotting the cues.

The strong emphasis on building a sense of community, igniting the passion and sense of ownership of their employees and bringing beauty to their customers' lives are aspects of engaging with the sacred. Organizations that include elements of the sacred in their worldview and their governing rules will likely be more successful in the 21st Century.

Make a Contribution

Lewin and Regine (2001) identify the sacred in organization with true felt actions that strengthen connections, enrich relationships and lead the organization as a whole to business success. They call this true felt action "care". They believe business built on good relationships and mutuality can be a source of social transformation. Engaging people with care engages the sacred at work by actualizing a higher purpose that serves the greater good. When people are fully engaged as human beings in the workplace, it generates the spirit of the organization.

To take this step and face the unknown and uncertainty requires an open and alert mind and heart. If we remember that we can influence the three conditions (containers, differences, exchange) that affect our relationship with the sacred, we can be open to surprises and the mysterious processes. Speaking about our fears and uncertainty will bring forth support from our connections with caring people.

Conclusion

Leaders are those people who decide to take initiative and interact in new ways to change the current patterns and culture of their groups. An adjacent possible for leaders in teams, organizations, and communities is the spiritual dimension of the groups they lead. Being a role model in practicing hospitality, respect, humility, dialogue, and other actions that demonstrate caring relationships is one way. Reexamining policies, procedures, goals and rituals to create more opportunities for meditation or prayer, or mental and physical exercise, is another. Creating more space for self-organizing to occur will allow members to speak their fears and uncertainties, surface what they are passionate about, and build a sense of community and what is sacred about their work and organization.

As organizations reassess their purpose by incorporating a greater recognition of the sacred and spirituality, they will tap into a power higher than a profit motive. The transformation needs to be at all levels in the organization – individually in the leaders and the employees, but also collectively in the policies and procedures.

Part IV The Sacred Awareness Journey

The sacred is everywhere, including the ambiguous spaces in our lives. Anywhere can be sacred. Developing a new sense of the sacred is a challenge faced by most persons of faith as well as for the non-religious. We need a sense of the sacred to create meaning and fill a spiritual void as we take in the implications of the new cosmology. Without a heightened sense of the sacred we may make decisions about our natural resources and our neighbors that will ultimately destroy life as we know it on earth.

Developing sacred awareness helps us navigate the dogmas and practices of both science and religion, and gives us the imagery and skills to live our lives forward. Clinging to "final answers" about what is sacred can provide a false sense of security and confidence that becomes a barrier to co-creation and adaptation to changing circumstances and environments.

Chapter 11 Aspects of the Journey

Morwood (2007) says the "everywhere God" is the mystery that holds everything in existence. How do we journey into that mystery, or, as Kauffman (2008) states it, how do we live our lives forward into the amazing creativity of the universe?

The Journey

The journey into the mystery and the creativity of the universe is a journey to sacred awareness, a journey to identify and engage with the sacred in our everyday life **(Figure 11-1)**.

Figure 11-1 A Compass for the Journey

This chapter will review the major themes discussed in the previous chapters in the five categories of the human life journey identified by Glenda Eoyang and the Human Systems Dynamics Institute (www.hsdinstitute.org).

- **Explore.** On the journey, we need to **explore** and identify the differences that are significant. What is it about our experience and views of the sacred that motivates us to engage with others about it?

- **Accommodate**. We also need to **accommodate** to our differences with others. Building constructive and meaningful exchanges with them will help adjust the differences.

- **Adapt**. Over time, we need to **adapt,** to establish new patterns and containers as the important differences shift.

- **Transform**. Eventually these new adaptive patterns will **transform** us and our relationships.

- **Co-evolve**. The new patterns will need to **co-evolve** as we respond to changes in our situation and environment as we continue to explore, accommodate, adapt, and transform.

Each aspect of the journey can be happening at any time with many different options for action. Early in the journey, we are likely to be emphasizing **exploring** and **accommodating** as new patterns are emerging. As we grow in sacred awareness, we will be more focused on **adaptation** and **transformation** of the new patterns. All along, we are **co-evolving** with others, our friends, our groups and organizations, and our communities.

Following is a brief description of the five aspects of the journey and some practical guides and simple rules for "working with ourselves as we are" and enhancing the presence of the sacred in our daily lives, in groups and organizations, and in our communities and society.

Explore

Young-Eisendrath & Miller (2000) define mature spirituality as acceptance of one's limitations, groundedness in the ordinary, and willingness to be surprised. They believe that neither divinity nor humanity can save us from ourselves. **We must work with ourselves as we are** – with blind spots and imperfections so that we can hone "integrity, wisdom and transcendence in the service of the question of what it means to be human in the Otherness of the universe".

The exploring aspect of the journey (which takes a lifetime) is daunting because it begins with ourselves. Our journey with the sacred requires us to become aware and mature about the creativity in the universe and the "God" that is everywhere. It requires us to see what is in the "bathwater" and what patterns we might influence to help bring about a different world.

Some guidelines for **explore:**

- Be patient about information that does not make sense as the new paradigm begins to form.

- Slow down; look deeper into what is happening.

- Stretch yourself to seek out and experience other cultures and religious traditions.

- Experience and embrace the new and confusing.

Accommodate

Can the human species evolve from self-centeredness to sacred-centeredness? Can we recognize the sacred in everyone and live in the world on that basis? Can we balance our competitiveness with cooperation? Can we become more inclusive? Can our thinking become more non-linear?

The accommodating aspect of the journey means we need to create the conditions for the sacred to be evident in us in a way that other people experience the sacred. Each of us has a great potential for self-empowerment and creating connections with others. We can envision what we want to do but we need to commit to the principles of sacred awareness that will unleash that potential.

We can expand our capacity to act. We are the agents of creation that can develop new patterns. Each of us has a gift needed by the whole. Connecting with others with a similar vision can transform our relationships, our groups and organizations, our communities, and ultimately our planet.

Some guidelines for **accommodate**:

- Make adjustments in our containers, the differences we identify as significant, and our exchanges with others.

- Let the patterns of a new more adaptive and sustainable sense of the sacred emerge.

- Help others make sense of what is happening to themselves and others.

- Fold and reflect on the learning from your new experiences.

Adapt

What we see as evil in the world has been described as things that take people away from the fullness of life, rejecting the sacred and thwarting the emergence of life, love, and what is possible. If we intentionally go to our adjacent possible we can avoid being complicit with the complex systems that cause evil and suffering.

We can adapt the insights from both theology and science in our understanding of the sacred. Spirit and creativity are everywhere in the universe – in us, our neighbor, and in the planet. Rather than be certain that we have the answer, we can open up to emergence, the surprises and unknowns, and live in relative comfort with that.

Raymo (2008) says that any religion worthy of humankind's future will be ecumenical, ecological, and embrace the scientific story of the world as the most **reliable** cosmology. The religion of the future will look for the signature of divinity in the wonder of the creation itself, not in supposed miracles or exceptions to nature's laws. Raymo quotes Ursula Goodenough saying:

> Emergence is inherent in everything that is alive, allowing our yearning for supernatural miracles to be subsumed by our joy in the countless miracles that surround us.

Some guidelines for **adapt**:

- Be flexible and willing to make adjustments as conditions change.

- Take risks -- make a decision and do something with incomplete information.

- Move into a balance between order and disorder, between certainty and uncertainty.

Transform

Morwood (2007) says that it is our responsibility to give the best possible human expression to the mystery of the sacred. How are we living what we profess? How do we relate with our 'neighbor'? How do we relate with our planet? How do we view ourselves? How do we make clear we are actively exhibiting 'the kingdom of God' in all we do and say? How do we express and renew our faith in the sacred? How do we get the power to do this?

Figure 11-2   Waking Up

We have been enamored with the power of technology to fix things. In the process we have lost our connection to each other and to the sacred mystery of the universe. We have been operating as if only we mattered without realizing the intricate connection we have to others and to the planet. The good news is that as we wake up **(Figure 11-2)** from the focus on technology and what we want to consume, we can wake up to exciting connections with others and the planet.

Some guides for **transform;**

- Foster transformation by creating the conditions for self-organizing (container, difference, exchange).

- Be clear about what actions to take.

- Be willing to learn from small experiments, realizing that experiencing the presence of the sacred gives passion and courage to individuals (Borg, 2006).

Co-evolve

Paul Hawken (2007), a researcher of groups involved in social change, says that a world-wide movement has emerged that is classless, unquenchable, and tireless. Its origins are the indigenous culture, the environment, and social justice movements. He says:

> This movement is humanity's immune response to resist and heal political disease, economic infection, and ecological corruption caused by ideologies. This is fundamentally a civil rights movement, a human rights movement; this is a democracy movement; it is the coming world.

Hawken says there are between 1 and 2 million organizations in the world, both secular and religious, who are working towards social and environmental justice and spiritual fulfillment. The actions of those involved may be the tipping point that will determine the kind of future we will pass on to our children and their children.

Some guides for **co-evolve:**

- Engage with your immediate situation and communicate that excitement to others.

- Provide support and compassion to others in their struggle with making changes.

Conclusion

Wendell Berry (2004), a philosopher, author, and farmer wrote:

> To live we must daily break the body and shed the blood of Creation. When we do this knowingly, skillfully, reverently, it is a sacrament. When we do it ignorantly, greedily, clumsily, destructively, it is a desecration. In such a desecration we condemn ourselves to spiritual and moral loneliness, and others to want.

Developing sacred awareness requires us to see the connectivity of everything and discern the part of the whole we are in a position to influence. All of these domains are connected. If we make a change in one domain, it will have a ripple effect. Using the concepts and methods suggested in this book will help us co-create with the sacred. We can join the sacred in manifesting the creativity in the universe in our relationships with those we love and our neighbor – and everyone is our neighbor.

Glossary of Methods and Metaphors

Adaptive

The ability of a system to change in response to challenges or opportunities presented by its internal or external environment

Adjacent Possible

Kauffman (2008) speculates that the idea of the adjacent possible is one of the general laws of the universe. He believes that, as an imperative, biological species keep expanding into what is adjacent and what is possible to jump to. In so doing they increase the diversity of what can happen next. However, the movement into the adjacent possible needs to be paced. To move too fast would destroy a species' own internal organization. In other words they "explore the adjacent possible as fast as they can get away with it".

Agents

The agents are the independent, interacting actors in a system.

Boundaries

Boundaries separate systems from their external environment and they separate internal sub-units from each other. It is where differences meet and generate learning and growth or conflict and disruption. Differences can be in demographics, expertise, discipline, role, power, location, etc.

Certainty Patterns

The Certainty Patterns are High Certainty, Self-Organizing, and Low Certainty. On a continuum of certainty-uncertainty they are patterns along the continuum, with more or less degree of certainty. The High Certainty pattern has a high degree of order and control. The Low Certainty pattern has a high degree of disorder and freedom. The Self-Organizing pattern has elements of both extremes.

Co-evolution

The evolution of two or more interdependent species, each adapting to changes in the other.

Complex Adaptive System (CAS)

A collection of semi-autonomous agents that create patterns by interacting in interdependent ways. These patterns reinforce the behavior of the agents.

Complexity Science

A new branch of science that studies the nonlinear aspects of the physical world.

Conditions of Self-Organization

Complex systems dynamics consist of patterns that are produced by the interaction of three conditions or forces – containing, differentiating, and exchanging. If any of the three conditions changes, the patterns change. Containers hold the system together. Differences differentiate the part from the whole. Exchanges involve how the parts contact and interact. We act on the patterns that we see. Behavior, emotions, and belief systems all have the three conditions embedded.

Constraints

Limitations that arise within a system that decreases the autonomy of the agents.

Container

One of the conditions that influences the speed, path, or direction of the process of self-organization. Containers hold the system together until patterns can form.

Differences

One of the conditions that influences the speed, path, or direction of the process of self-organization. Differences provide the potential for change in a system. It can refer to the variance in one variable or multiple variables in the system.

Downward and Upward Causation

In upward causation, the parts create the whole. In downward causation, the emergent properties affect the existence, properties, and interaction of the parts. For example, our eating, drinking, and exercise behavior affect the health of our body's cells, just as our cells affect our well-being.

Emergence

Higher levels of structure, dynamic interaction, and system properties are generated by a complex set of causes. These higher levels are emergent forms of synergy that are more than the sum of the parts.

Entropy

In Newtonian physics, systems tend toward randomness and disorder and to run down unless they are renewed.

Equifinality

The ability of a system to reach the same end in many different ways

Exchanges

One of the conditions that influences the speed, path, or direction of the process of self-organization. Exchanges that connect parts of the system to each other and to the environment to share information and other resources can transform the system.

Feedback Loops

Feedback occurs when output from an event in the past will influence the same event in the present or future. When an event is part of a chain of that forms a loop, then the event is said to "feed back" into itself. A feedback loop is the causal path that leads from the initial interaction to the subsequent modification of the event. Feedback loops vary by length, width, and dynamic.

Fractal Patterns

Any pattern repeated at different levels throughout the system. The shapes are similar at all scales (e.g. broccoli). Biological scaling gives coherence in widely diverse entities, e.g. oak tree. Recognizing patterns of basic values or simple rules generates diverse, but self-similar behavior across scales. Naming and telling stories about dynamics in a system help reinforce and shape fractal patterns, e.g. cultural norms, leadership behaviors. Subgroups or individuals mimic the overall behavior in the larger system.

Homeostasis

The tendency of a system to avoid changing.

Human Systems Dynamics

Emergent field of study and practice developed by Glenda Eoyang that explores how humans live, work and play together.

Influence Factors

Both as individuals and as members of groups we are influenced by three factors. When we have a decision to make, these multiple patterns are working: worldview, rules, and reality. The conditions of self-organizing also influence each of these factors.

Integration/Differentiation

The tension between the system pressure for all sub-units to be the same, and the sub-units' push to be unique and different

Nonlinear

The outputs of a complex system are not proportional to their inputs. The parts interfere with each other so that the system can only be analyzed in a holistic way.

Nonergodic universe

The universe is incomprehensible in certain crucial respects. The transitions are so variable there are not enough observations to ascertain the probabilities.

Non-reductionist science

The scientific assumptions about linearity, order, uniformity, and hierarchy (natural laws) do not explain everything. Reductionist science can't explain novelty or human consciousness that is grasped by truth, unity, being, and beauty. Kauffman (2008) replaces reductionism with the non-reductionist sciences of emergence and complexity to explain what happens when conditions are nonlinear and probabilistic.

Organic/autopoietic systems

Self-maintaining systems that are adaptive to their environment.

Pantheism

Pantheism says that God and the universe are coextensive, that God is everything.

Panentheism

Panentheism holds that everything in the universe is contained within God, but God is also greater than the universe.

Patterns

Similarities, differences and relationships that have meaning across space and/or time.

Sacred

The sacred is the creativity in the universe (a complexity science perspective). The sacred is God's presence in creation and in us (theological perspective).

Sacred Awareness

Sacred awareness is the capacity for recognizing and engaging the sacred in orienting our daily lives. This capacity allows us to examine what is important to us about the sacred, including all of life and the planet; understand the information we use in interacting with others about the sacred; and increase our awareness of the assumptions we make about the sacred that either perpetuates our separation or enhances our connection to others.

Self-Organized Criticality

Nonlinear dynamical systems exhibit order, criticality, and chaos. Change occurs when there is a balance between order and chaos (edge of chaos). The theory of self-organized criticality may provide laws for the evolution on higher levels of the biosphere, despite absence of law at lower levels. Complex human systems have diverse and ever-changing subsystems and processes that fit with one another and with larger systems more or less seamlessly. They exist for periods of time, and then evolve further. For example, the economy may coevolve into its adjacent possible in a self-organized critical way (Kauffman, 2008).

Self-Organization

Patterns are generated as agents in a system interact and form feedback loops. Relationships within the system create order. A system that is pushed far-from-equilibrium restructures itself. People will organize themselves – this can be good or bad. You can influence, but not control, self-organizing processes. The amount of time it takes to self-organize is unpredictable.

Semi-autonomous

Agents within a system have a degree of freedom of choice within the system, but they also are constrained by being part of the system.

Sensitive Dependence on Initial Conditions

Small causes can have a large effect in systems if the initial conditions are "far from equilibrium" and unstable. The phenomenon is also known as the "butterfly effect", a metaphor for the theoretical possibility that a small disturbance in the air currents off the coast of Africa, if the conditions are right, can morph into a hurricane that could hit Central and North America. The reverse is also true. A large disturbance can have little impact if the initial conditions are not conducive to system change.

Simple Rules

If the members of a system all follow the same short list of simple rules, then as a whole the system behaves in a coherent way. If the rules contradict each other, it means the system will exist with ambiguity

Stretch and Fold (Baker's Transformation)

To establish diverse yet coherent systems, we stretch to add differences, amplify current differences, expand our range/scope, and include new things. We fold to look for common ground, move into small time and/or space, build tight connections, and clarify purposes and objectives.

References

Bessler-Northcutt, J. (2004). "Learning to see God: Prayer and practice in the wake of the Jesus seminar. In Jesus Seminar, *The historical Jesus goes to church*. Santa Rosa, CA: Polebridge Press, pp. 51-63.

Beutner, E.F. (Nov-Dec., 2004). The burglar who makes appointments. *The Fourth R*, pp. 17-18.

Berry, W. (2004). The unsettling of America: Culture & agriculture. Sierra Club Books

Bohm, D. (2002). *Wholeness and the implicate order*. London: Routledge

Borg, M. (2003). *The heart of Christianity*. San Francisco: Harper.

Cannato, J. (2006). *Radical Amazement: Contemplative lessons from black holes, supernovas, and other wonders of the universe*. Notre Dame: Sorin Books.

Chopra, D. (2000). *How to know God: The soul's journey into the mystery of mysteries*. New York: Three Rivers Press.

Clayton, P. (Jan 13, 2004a) *Emerging God: Theology for a complex universe*, Christian Century.

Clayton, P. (2004b). *Mind and emergence: From quantum to consciousness*. Oxford University Press.

Clayton, P. & Davies, P. (2006). *The re-emergence of emergence: The emergentist hypothesis from science to religion*. N.Y.: Oxford University Press.

Cobb, J.D. Jr. (2007) *A Christian natural theology*, 2nd ed. John Knox Press.

Collins, F. (2006). *The language of God: A scientist presents evidence for belief.*

Conforti, M. (2003). *Field, form and fate. Patterns in mind, nature, and psyche.* Spring Journal, Inc.

Davies, P. (2004). Teleology without teleology: Purpose through emergent complexity, in Clayton, P. and Peacocke, A., eds., *op. cit.* pp. 95-108.

Davies, P. (2007). The *cosmic jackpot: Why our universe is just right for life*. N.Y. Houghton-Mifflin.

Deloria, Jr. V. (1994). *God is red: A native view of religion*. Golden, Colorado: Fulcrum Publishing.

Denis E. (2006). *Ecology at the heart of faith*. Maryknoll, N.Y.: Orbis Books

Dowd, Michael (2008). *Thank God for evolution: How the marriage of science and religion will transform your life and our world.* N.Y.: Viking.

Eoyang, G. (1999). *Coping with chaos: Seven simple tools*. Cheyenne: Lagumo.

Fox, M. (2006). *A new reformation: Creation spirituality and the transformation of Christianity.* Rochester, Vermont: Inner Traditions

Gingerich, O. (2000), "Is there design and purpose in the universe?" in Haught, J. F, ed. *Science and religion in search of cosmic purpose.* Georgetown University Press, pp. 121-132

Goldstein, J. (1994). *The unshackled organization*. New York: Productivity Press.

Greene, B. (2005). *The fabric of the cosmos: Space, time, and the texture of reality.* Knopf.

Handy, C. (1998). *The hungry spirit: Beyond capitalism: A quest for purpose in the modern world.* N.Y.: Broadway Books.

Hawken, P. (2007). *Blessed unrest: How the largest movement in the world came into being and why no one saw it coming.* Viking.

Heschel, A.J. (1976). *God in search of man: A philosophy of Judaism.* Farrar, Strauss & Giroux.

Human Systems Dynamics Institute. Circle Pine, MN www.hsdinstitute.com

Idliby, R., S Oliver, & P. Warner. (2007). *The faith club: A Muslim, a Christian, a Jew – three women search for understanding.* New York: The Free Press.

Jeeves, M., ed. (2004). *From cells to souls – and beyond: Changing portraits of human nature.* Eerdmans.

Johnson, E.A. (2007). *Quest for the living God: Mapping frontiers in the theology of God.* New York: Continuum.

Jones, R. (2005). *God, Galileo & Geering: A faith for the 21st century.* Polebridge Press.

Jung, C.G. (1989). *Memories, dreams, reflections*. Vintage

Kaufman, G. D. (2004). *In the beginning…creativity*. Minneapolis: Fortress Press.

Kauffman, S.A. (2008). *Reinventing the sacred: A new view of science, reason, and religion.* New York: Basic Books.

Kelly, S. & Allison, M.A. (1998). *The complexity advantage: How the science of complexity can help your business achieve peak performance.* New York: McGraw-Hill.

Lewes, G. H. (1875), written at London, *Problems of Life and Mind (First Series)*, vol. 2, Trübner

Lewin, R & Regine, B. (2001). *Weaving complexity & business: Engaging the soul at work.* New York: Texere.

Machen, J. G. (1978). *The Christian faith in the modern world.* Eerdmans Publishing Company.

Macnab, F. (2004). Preaching the new faith, in Hoover, R.H., et al, The historical Jesus goes to church, op. cit., pp. 105-121.

Morwood, M (2004). *Praying a new story.* Maryknoll, N.Y.: Orbis Books

_____ (2007). *From sand to solid ground: Questions of faith for modern Catholics.* New York: Crossroads Publishing

Nasr, S. H. "God: The reality to serve, love, and know", in M. Borg & R. Mackensie, eds. (2000). *God at 2000.* Harrisburg, Pennsylvania: Morehouse Publishing, pp. 99-134.

Nelson, D.W. Rabbi (2005). *Judaism, physics, and God: Searching for sacred metaphors in a post-Einstein world.* Woodstock, Vermont: Jewish Lights Publishing.

Olson, E. E. & Eoyang, G. H. (2001). *Facilitating organization change: Lessons from complexity science.* San Francisco: Jossey-Bass/Pfeiffer.

Pachamama Alliance San Francisco, CA. www.pachamama.org

Padgett, A.G. (2007) Encyclopedia of Philosophy, 2nd edition. Macmillan/Gale Group.

Peters, K. E. (2008). *Spiritual transformations: Science, religion, and human becoming.* Minneapolis: Fortress.

Petzinger, T. (1999). *The new pioneers.* New York: Simon and Schuster.

Polkinghorne, J. (2005). *Exploring reality: The intertwining of science and religion. New Haven:* Yale University Press.

Raymo, Chet (2008). *When God is gone everything is holy.* Sorin Books.

Sacks, J.S. (Oct 7, 2007). Religion and science are twin beacons of humanity. *The Times*. http://www.timesonline.co.uk/tol/comment/faith/article2607585.ece?print=yes&randnum =

Senge, P. et. al (2005). *Presence: Human purpose and the field of the future*. Broadway Books.

Spong, J. S. (2007). *Jesus for the non-religious; Rediscovering the divine at the heart of the human*. San Francisco: Harper.

Stacey, R., Griffin, D. &. Shaw, P. (2000). *Complexity and management: Fad or radical challenge to systems thinking?*. London: Routledge.

Swimme, B. (1999). *The hidden heart of the cosmos: Humanity and the new story*. Orbis Books.

Wessels, C. (2003). *Jesus in the new universe story*. Orbis Books.

Williams, P.A. (2008a). *Revealing God: A new theology from science and Jesus*. Infinity Publishing.com

_____ (2008b). Science and the bible in dialogue III: In the end: Biology. *The Fourth R,* vol. 21, no. 6, pp. 15 – 21.

Young-Eisendrath, P. & M.E. Miller (2000). *The Psychology of mature spirituality: Integrity, wisdom, transcendence*. Philadelphia: Routledge.

Zimmerman, B., Lindberg, C. & Plsek, P. (1998). *Edgeware: Insights from complexity science for health care leaders*. Irving TX: VHA, Inc.

About the Author

Ed Olson is an applied behavioral scientist living in Longville, Minnesota and Estero, Florida. He has led organization change, team building, management development, and workforce diversity initiatives for Texaco, Inc., U.S. General Accounting Office, Digital Equipment Corporation, NASA, and the Food and Drug Administration. He facilitates human interaction, group dynamics, and complex system change workshops. He leads workshops on the Singer-Loomis Type Deployment Inventory and the Awakening the Dreamer symposium.

He is a collegiate professor at the University of Maryland, University College Graduate School and an adjunct professor in the Executive Leadership Program, the George Washington University. Formerly he was professor of management at Baldwin-Wallace College (Berea, Ohio) and professor of information service at the University of Maryland, College Park. While at Maryland he co-founded the Maryland Center for Productivity and Quality of Work Life and the Human Relations Processes program.

His research interests are in applications of complexity theory and analytical psychology. Ed is the co-author of *Facilitating Organization Change: Lessons from Complexity Science* (Jossey-Bass/Pfeiffer, 2001).

He holds a B.A. magna cum laude (Philosophy) from St. Olaf College, M.S. (Pastoral Counseling) from Loyola College (Baltimore), and an M.A. and Ph. D (Government and Public Administration) from the American University. Ed is a professional member of the NTL Institute for Applied Behavioral Science, an associate of the Human Systems Dynamics Institute, a presenter for the Pachamama Alliance, and member of the Organization Development Network, the C.G. Jung Institute of New York, and Phi Beta Kappa. He serves on the Board of the Center for Sacred Unity at the Lamb of God Church in Estero, Florida.

Appendix A   Conditions of Self-Organizing and the Trinity

A sense of involvement with the sacred has changed along with the culture and experience of its believers. Spong (2006) points out that every human attempt to define God is nothing more than our definition of our experience of the divine. We can never tell who God is or who God is not. In this appendix I compare the Conditions of Self-Organizing with several perspectives on a divine trinity as an example of how the rules of self-organizing are developed from the experience of faith communities (**See Figure 1**). If the divine created the universe's self-organizing processes and the three simple rules of the model, how can we comprehend this? Polkinghorne (2005) believes that Trinitarian ways of thought are consistent with the kind of relations and holistic thinking about the source of created reality expressed in the new sciences.

Early Christian Communities.

- **Containing.** The early Christian communities named the containing forces as **God** the creator. As the ultimate container, we can think of the God who created the laws that generated the universe. Purpose and meaning come from God. The boundaries of the multiple containing processes that God set forth are permeable and indeterminate.

- **Differentiating.** As Jesus dramatically brought the differentiating force he became the **Christ**, the source of what is significant and what must be attended to in relation to God. Jesus shocked the domination systems by raising issues of social justice and resetting priorities. As an advocate and counselor Jesus identified what is important – to forgive and be forgiven, to rescue the less fortunate, and to seek justice. Jesus invites us to experience a new humanity, an inclusive humanity expanded beyond tribal boundaries, prejudices, and insecurities and beyond the quest for survival (Spong, 2006).

- **Transforming Exchanges.** The transforming exchanges between agents of creation and between those agents and God and Christ were seen as the **Holy Spirit**. The Holy Spirit reveals God and Christ and connects us to each other in ways that are transforming. We find joy and companionship in nurturing and comforting exchanges with the human family and the cosmos. Edwards (2004) describes the spirit as the breath of God breathing life into a universe of creatures and the source of the new in an emergent universe. Through the indwelling spirit, creatures of the universe are brought into communion with one another.

The Containing function of the Trinity cannot be solely assigned to one person. Several recent authors in the field of science and religion have described the Trinity as entangled with itself as well as with the creation it made possible (Polkinghorne, 2005; Bonting, 2005; Peacocke, 2003; Jeeves, 2004; Welker, 2004). Bonting, for example, believes God the Father sends the energy needed for initial creation; the Holy Spirit infuses the information required for continuing creation; the cosmic Christ completes and redeems creation and provides the energy for ongoing creation. Simmons (2006) describes the entanglement of the inner relationship of the three persons of the Trinity as a dance.

- **Containing.** Adopting the right container (God) lets the transforming exchange (Spirit) fill us and we will find the courage to do what is important (Christ). However, the Bible references Christ as holding everything together and the Spirit, in effect, creates new containing processes as people are brought together in relationship with each other and God.

- **Differentiating.** Significant difference (Christ) brings about the transforming exchanges (Spirit) and the grounding (God) for our work. However, the differentiating function is also not limited to Christ. God created all of the significant differences in the Universe and the Holy Spirit stirs emotions and shifts our focus.

- **Exchange.** Transforming exchange (Spirit) helps us to discern a significant Difference (Christ) while the Containing process (God) keeps us balanced. However, God's covenant of creation serves a transforming function by connecting us to God and each other. Christ connects us with God through his work of reconciliation and examples of obedience to God.

Spong's non-theistic Images. Spong (April, 2007) offers three non-theistic images of God:

- **Containing.** God is rock underneath us. It is the trustworthy form that holds us. God as the ground of our being comes when we become who we can be, absorbing the crises of life without falling apart.

- **Differentiating.** God is love – a life of power. We can't create it; we can give it; our bodies are channels through which the transcendent power of love flow and call us into a deeper level of humanity. The more we love, the more God is present.

- **Exchange.** God is wind and the breath within us – power that flows through every living thing. The breath of God produces life and vitalizes all that exists. We worship God by living fully and; as we do, God is more evident.

Geering's Secular Trinity. Geering in Jones (2005) believes a secular trinity is now evident:

- **Containing** - an evolving universe with all its mystery and power.

- **Differentiating** - human culture and creativity at the vanguard of the evolutionary process.

- **Exchange** - an emerging global consciousness that has the potential to bring humankind into a community of peace and good will.

Other Religions

The history and sacred writings of all religions could be looked at from this perspective to discern when and under what conditions the divine has been present in creation. From the Judaic tradition, Rabbi Nelson (2005) suggests new metaphors for God – Big Bang and gravity (Container); light (Difference); and fractals (Exchange). In Hinduism, according to Oomen (2002) the Vedantic understanding of Brahman would suggest that the containing force is *sat* (being, the still God); the differentiating force is *sit* (knowledge and the journeying God); and the transforming force is *Ananda* (Joy, the returning God).

Figure 1

Characteristics of Trinitarian Forces

Christian Trinity	Spong	Geering	Conditions of Self-organizing
First Person of the Trinity			
God creates multiple containers of meaning, being, gives purpose (covenant), creates and destroys, sets indeterminate and determinate boundaries	God as rock underneath us	Evolving universe with all its mystery and power	**Containing** – boundaries are created for the self-organizing process
Second Person of the Trinity			
Christ sets priorities, shocks the system, reconciles, mediates, rescues, demands justice, forgives, advocates, counsels	God as love – a life of power	Human culture and creativity at the vanguard of the evolutionary process	**Differentiating** – significant differences focus and guide the self-organizing process
Third Person of the Trinity			
The Holy Spirit is a companion that reveals, exchanges, comforts, sanctifies.	God as wind and the breath within us	Emerging global consciousness with potential to bring humankind into a community of peace and good will	**Transforming** exchanges – interactions across differences create new patterns

**The Center for Sacred Unity
at Lamb of God Church**

*Embracing an Emerging world, an Emerging Church, an
Emerging Soul and an Emerging Cosmology for
21ˢᵗ Century Christians*

Center for Sacred Unity

Mission Statement

BUILDING ON OUR CHRISTIAN TRADITION EMPOWERING AND EQUIPPING PEOPLE OF FAITH FOR LIVES OF SERVICE, MISSION AND MINISTRY IN THE 21ST CENTURY

Our Task

- Affirming and celebrating the diversity and mystery of God

- Building and sustaining working relationships with other sacred traditions to bless the world together

- Developing and teaching models for Christian discipleship for 21st Century Christians

- Embracing an emerging world, an emerging Church, an emerging Soul, and an emerging Cosmology

- Bridging the evolving relationship between science and theology

*   *   *   *   *   *   *   *   *   *   *

Keep the Bathwater: Emergence of the Sacred in Science & Religion

Available from:

Center for Sacred Unity
Lamb of God Church
19691 Cypress View Drive
Fort Myers, FL 33967

Contact Pastor Walter Fohs
Fax: 239-267-3043   Phone: 239-267-3525
Email: office@LambofGodChurch.net